# AIM for Success Practice Sheets

# Beginning & Intermediate Algebra

**THIRD EDITION**

**Richard N. Aufmann**
Palomar College

**Joanne S. Lockwood**
Nashua Community College



Prepared by

**Christine S. Verity**

BROOKS/COLE
CENGAGE Learning

Australia • Brazil • Japan • Korea • Mexico • Singapore • Spain • United Kingdom • United States

BROOKS/COLE
CENGAGE Learning



ISBN-13: 978-1-133-36377-4
ISBN-10: 1-133-36377-6

**Brooks/Cole**
20 Davis Drive
Belmont, CA 94002-3098
USA

Cengage Learning is a leading provider of customized learning solutions with office locations around the globe, including Singapore, the United Kingdom, Australia, Mexico, Brazil, and Japan. Locate your local office at: **www.cengage.com/global**

Cengage Learning products are represented in Canada by Nelson Education, Ltd.

To learn more about Brooks/Cole, visit **www.cengage.com/brookscole**

Purchase any of our products at your local college store or at our preferred online store **www.cengagebrain.com**

Printed in the United States of America
1 2 3 4 5 6 7 16 15 14 13 12

# Contents

Name Score

**Place the correct symbol, < or >, between the values of the two numbers.**

**1.** –1 –3

**2.** –8 –5

**3.** 7 0

**4.** –24 42

**5.** 53 –64

**6.** –55 53

**1.** __________
**2.** __________
**3.** __________
**4.** __________
**5.** __________
**6.** __________

**Answer True or False.**

**7.** $-5 > -2$

**8.** $3 < -5$

**9.** $-12 < 14$

**10.** $-34 > 47$

**11.** $19 < -31$

**12.** $-10 > -7$

**13.** $53 < -71$

**14.** $-53 < -81$

**15.** $84 > 73$

**7.** __________
**8.** __________
**9.** __________
**10.** __________
**11.** __________
**12.** __________
**13.** __________
**14.** __________
**15.** __________

**Use the roster method to write the set.**

**16.** The natural numbers less than 4

**17.** The negative integers greater than or equal to –3

**18.** The positive integers less than 6

**19.** Given $A = \{-6, -3, 1, 4\}$, which elements of set $A$ are less than –1?

**20.** Given $B = \{-9, -5, -1, 3\}$, which elements of set $B$ are greater than –6?

**21.** Given $C = \{-20, 0, 1, 18\}$, which elements of set $C$ are less than 1?

**22.** Given $D = \{-16, -5, 2, 3, 7, 18\}$, which elements of set $D$ are greater than or equal to 2?

**23.** Given $E = \{-26, -19, -7, -5, 2, 6, 9\}$, which elements of set $E$ are less than or equal to 6?

**24.** Given $F = \{-35, -15, -1, 0, 3, 6\}$, which elements of set $F$ are greater than –1?

**16.** __________
**17.** __________
**18.** __________
**19.** __________
**20.** __________
**21.** __________
**22.** __________
**23.** __________
**24.** __________

Name Score

**Find the additive inverse.**

**1.** 7

**2.** –4

**3.** –5

**4.** 34

**5.** –28

**6.** 66

**Evaluate.**

**7.** $(-3)$

**8.** $-(7)$

**9.** $-(-5)$

**10.** $-(-13)$

**11.** $-|-4|$

**12.** $|15|$

**13.** $|-16|$

**14.** $-|19|$

**15.** $-|-19|$

**Place the correct symbol, < or >, between the values of the two numbers.**

**16.** $-|21| \quad |-26|$

**17.** $-|22| \quad -|31|$

**18.** $-|14| \quad -|-35|$

**19.** $-|-33| \quad |54|$

**20.** $|40| \quad |-54|$

**21.** $-|28| \quad |9|$

**22.** $|6| \quad -|43|$

**23.** $-|-25| \quad |18|$

**24.** $|-15| \quad |4|$

1. ______________
2. ______________
3. ______________
4. ______________
5. ______________
6. ______________
7. ______________
8. ______________
9. ______________
10. ______________
11. ______________
12. ______________
13. ______________
14. ______________
15. ______________
16. ______________
17. ______________
18. ______________
19. ______________
20. ______________
21. ______________
22. ______________
23. ______________
24. ______________

Name Score

**Add.**

**1.** $2 + (-4)$

**2.** $-5 + 3$

**3.** $-3 + (-7)$

**4.** $-16 + 23$

**5.** $9 + (-7)$

**6.** $-10 + (-8)$

**7.** $-1 + (-6)$

**8.** $-7 + (-7)$

**9.** $5 + (-10)$

**10.** $-9 + 3$

**11.** $-7 + 11$

**12.** $-13 + 9$

**13.** $6 + (-4) + (-7)$

**14.** $-4 + (-10) + (-11)$

**15.** $11 + (-3) + (-13)$

**16.** $-15 + (-5) + 30$

**17.** $10 + 54 + (-28)$

**18.** $-4 + (-9) + 9$

**19.** $-36 + (-40) + (-16)$

**20.** $11 + (-20) + (-30)$

**21.** $-11 + (-6) + 8$

**22.** $16 + (-18) + 5 + (-4)$

**23.** $-12 + (-4) + 6 + (-7)$

**24.** $-23 + 9 + 3 + (-16)$

**25.** $6 + 8 + (-13) + (-4)$

**26.** $-12 + (15) + (-16) + (-10)$

**27.** $-21 + (-33) + 27 + 18$

**1.** ______________
**2.** ______________
**3.** ______________
**4.** ______________
**5.** ______________
**6.** ______________
**7.** ______________
**8.** ______________
**9.** ______________
**10.** ______________
**11.** ______________
**12.** ______________
**13.** ______________
**14.** ______________
**15.** ______________
**16.** ______________
**17.** ______________
**18.** ______________
**19.** ______________
**20.** ______________
**21.** ______________
**22.** ______________
**23.** ______________
**24.** ______________
**25.** ______________
**26.** ______________
**27.** ______________

Name Score

**Subtract.**

| | | |
|---|---|---|
| **1.** 13 – 7 | **2.** 15 – 5 | **3.** 5 – 15 |
| **4.** –6 – 3 | **5.** –7 – 5 | **6.** 6 – (–4) |
| **7.** 2 – (–2) | **8.** –9 – (–1) | **9.** –3 – (–7) |
| **10.** –17 – 11 | **11.** –5 – 2 – 1 | **12.** –4 – 5 – 11 |
| **13.** 4 – 5 – 6 | **14.** 10 – (–9) – 7 | **15.** –3 – 10 – (–6) |
| **16.** 12 – 9 – 11 | **17.** –4 – (–5) – (–7) | **18.** 6 – 7 – (–2) |
| **19.** –28 – (–52) – 24 – 9 | **20.** 44 – (–80) – 60 – 10 | **21.** –17 – 36 – 62 – 14 |
| **22.** 38 – (–21) – 62 – (–17) | **23.** –36 – (–58) – 10 – 17 | **24.** –16 – 37 – (–64) – 35 |
| **25.** –19 – (–31) – (–76) – 24 | **26.** 21 – 33 – 57 – 14 | **27.** 38 – (–22) – 63 – (–9) |

**1.** ____________
**2.** ____________
**3.** ____________
**4.** ____________
**5.** ____________
**6.** ____________
**7.** ____________
**8.** ____________
**9.** ____________
**10.** ____________
**11.** ____________
**12.** ____________
**13.** ____________
**14.** ____________
**15.** ____________
**16.** ____________
**17.** ____________
**18.** ____________
**19.** ____________
**20.** ____________
**21.** ____________
**22.** ____________
**23.** ____________
**24.** ____________
**25.** ____________
**26.** ____________
**27.** ____________

Name Score

**Multiply.**

**1.** $11 \times 4$ **2.** $-2 \times 4$ **3.** $3(-4)$

**4.** $-7(-2)$ **5.** $(8)(-2)$ **6.** $4(-6)$

**7.** $(-6)(7)$ **8.** $(-4)(5)$ **9.** $(-4)(-7)$

**10.** $(-2)(-9)$ **11.** $(0)(-6)$ **12.** $-21 \times 3$

**13.** $18(-5)$ **14.** $3(-19)$ **15.** $-4(-13)$

**16.** $-4(27)$ **17.** $9(-18)$ **18.** $-6(-32)$

**19.** $-5(36)$ **20.** $3(-7)(2)$ **21.** $(-3)(5)(8)$

**22.** $6 \times 6 \times (-3)$ **23.** $(-8)(3)(-2)$ **24.** $-7(-6)(-5)$

**25.** $-4(7)(-2)$ **26.** $(-5)(4)(9)$ **27.** $4(-3)(4)$

**1.** ______
**2.** ______
**3.** ______
**4.** ______
**5.** ______
**6.** ______
**7.** ______
**8.** ______
**9.** ______
**10.** ______
**11.** ______
**12.** ______
**13.** ______
**14.** ______
**15.** ______
**16.** ______
**17.** ______
**18.** ______
**19.** ______
**20.** ______
**21.** ______
**22.** ______
**23.** ______
**24.** ______
**25.** ______
**26.** ______
**27.** ______

Name Score

**Divide.**

**1.** $18 \div (-6)$

**2.** $(-36) \div (6)$

**3.** $(-32) \div (-4)$

**4.** $7 \div (-7)$

**5.** $-42 \div 6$

**6.** $40 \div (-8)$

**7.** $-30 \div 5$

**8.** $-48 \div 6$

**9.** $-72 \div (-8)$

**10.** $-42 \div (-7)$

**11.** $-42 \div (-6)$

**12.** $-50 \div 5$

**13.** $-55 \div (-11)$

**14.** $-84 \div (-7)$

**15.** $-60 \div (-5)$

**16.** $-75 \div (-5)$

**17.** $105 \div 7$

**18.** $68 \div (-4)$

**19.** $72 \div (-9)$

**20.** $(-128) \div 8$

**21.** $192 \div (-12)$

**22.** $154 \div (-22)$

**23.** $-182 \div (-13)$

**24.** $205 \div (-5)$

**25.** $-120 \div (-8)$

**26.** $135 \div 15$

**27.** $168 \div (-12)$

1. ______
2. ______
3. ______
4. ______
5. ______
6. ______
7. ______
8. ______
9. ______
10. ______
11. ______
12. ______
13. ______
14. ______
15. ______
16. ______
17. ______
18. ______
19. ______
20. ______
21. ______
22. ______
23. ______
24. ______
25. ______
26. ______
27. ______

Name Score

**Solve.**

**1.** The highest elevation in Asia is Mt. Everest with an elevation of 8850 m. The lowest point is the Dead Sea with an elevation of –400 m. What is the difference between the highest and lowest points in Asia?

**2.** The average high temperature in Fargo, ND during the month of July is 89.6°F. The average low temperature in February is –17.6°F. Find the difference between the two average temperatures.

**1.** ____________

**2.** ____________

**3.** When a certain amount of salt is added to water, the freezing point of the salt water mixture is –4°C. The boiling point of the mixture is 103°C. What is the difference between the boiling point and the freezing point of the salt water mixture?

**4.** The temperature at which a rare element boils is 350°C. This element freezes at –29°C. Determine the difference between the temperature at which it boils and the temperature at which it freezes.

**3.** ____________

**4.** ____________

**5.** When the temperature is –10°F and the wind is blowing at 15 mph, the wind chill factor is –32°F. For the same temperature of –10°F and the wind blowing at 30 mph, the wind chill factor is –39°F. Find the difference between the wind chill factors.

**6.** The temperature at which Xenon boils is –107°C. Xenon melts at –112°C. Determine the difference between the temperatures at the boiling point and the melting point.

**5.** ____________

**6.** ____________

**7.** The daily low temperatures during one week were recorded as follows: 4°, –6°, 8°, –2°, –9°, –11°, and –5°. Find the average daily low temperature for the week.

**8.** The daily high temperatures during one week were recorded as follows: –5°, –8°, 6°, 8°, 0°, –6°, and –2°. Find the average daily high temperature for the week.

**7.** ____________

**8.** ____________

**9.** One golfer had a score of six under par (–6) while a second golfer had a score of 8 over par (+8). Find the difference between their scores.

**10.** A four-member golf team had a combined score of 18 under par (–18). Another team had a combined score of 2 over par (+2). Find the difference between their scores.

**9.** ____________

**10.** ____________

Name Score

**Write as a decimal. Place a bar over repeating digits.**

**1.** $\frac{3}{8}$

**2.** $\frac{7}{9}$

**3.** $\frac{1}{12}$

**4.** $\frac{7}{15}$

**5.** $\frac{5}{16}$

**6.** $\frac{3}{20}$

**7.** $\frac{9}{20}$

**8.** $\frac{19}{20}$

**9.** $\frac{3}{25}$

**10.** $\frac{7}{25}$

**11.** $\frac{11}{30}$

**12.** $\frac{15}{30}$

**1.** ____________

**2.** ____________

**3.** ____________

**4.** ____________

**5.** ____________

**6.** ____________

**7.** ____________

**8.** ____________

**9.** ____________

**10.** ____________

**11.** ____________

**12.** ____________

Name Score

**Simplify.**

1. $\frac{1}{3}\times\left(-\frac{3}{5}\right)$

2. $-\frac{5}{6}\times\left(-\frac{9}{20}\right)$

3. $\left(-\frac{4}{15}\right)\left(-\frac{3}{16}\right)$

4. $\left(\frac{3}{8}\right)\left(\frac{9}{10}\right)\left(-\frac{8}{15}\right)$

5. $-\frac{6}{7}\cdot\left(-\frac{15}{22}\right)\left(\frac{28}{35}\right)$

6. $\left(-\frac{3}{4}\right)\left(\frac{12}{25}\right)\left(\frac{20}{27}\right)$

7. $1.5\times2.84$

8. $-0.6\times5.4$

9. $(-1.78)(-1.2)$

10. $\frac{5}{9}\div\frac{1}{3}$

11. $\frac{9}{16}\div\left(-\frac{3}{8}\right)$

12. $-\frac{7}{10}\div\frac{21}{25}$

13. $-\frac{35}{48}\div\frac{7}{20}$

14. $\frac{21}{32}\div\frac{7}{12}$

15. $-\frac{9}{22}\div\frac{6}{11}$

16. $\frac{5}{12}\div\left(-\frac{25}{36}\right)$

17. $-\frac{8}{21}\div\left(-\frac{2}{7}\right)$

18. $-\frac{8}{9}\div\left(-\frac{2}{3}\right)$

19. $-\frac{5}{12}\div5$

20. $-0.918\div(-0.204)$

21. $\left(-\frac{3}{7}\right)\div\left(-\frac{14}{3}\right)$

22. $6.15\div(-37.8)$
Round to the nearest thousandth.

23. $-0.2354\div(-17.45)$
Round to the nearest thousandth.

24. $-\frac{2}{3}\div\frac{4}{7}-\frac{3}{4}$

1. ______
2. ______
3. ______
4. ______
5. ______
6. ______
7. ______
8. ______
9. ______
10. ______
11. ______
12. ______
13. ______
14. ______
15. ______
16. ______
17. ______
18. ______
19. ______
20. ______
21. ______
22. ______
23. ______
24. ______

Name Score

**Simplify.**

1. $-\frac{3}{10}+\frac{1}{4}$

2. $-\frac{3}{5}+\frac{11}{20}$

3. $-\frac{5}{12}+\frac{11}{24}$

4. $-\frac{8}{9}+\frac{8}{15}$

5. $-\frac{7}{16}+\left(-\frac{3}{4}\right)+\frac{5}{8}$

6. $\frac{3}{8}+\left(-\frac{7}{12}\right)+\frac{1}{4}$

7. $-\frac{5}{9}+\frac{8}{15}+\frac{4}{5}$

8. $-9.037 + 5.7$

9. $3.27 + 5.816$

10. $\frac{1}{6}+\frac{2}{9}+\frac{5}{18}$

11. $-\frac{7}{8}+\left(-\frac{7}{12}\right)$

12. $-7.63 + 32.49 + (-89.35)$

13. $\frac{3}{4}-\frac{1}{6}$

14. $\frac{5}{6}-\frac{7}{18}$

15. $-\frac{7}{24}-\frac{5}{16}$

16. $-\frac{5}{18}-\frac{5}{6}$

17. $-\frac{3}{8}-\left(-\frac{5}{6}\right)$

18. $\frac{2}{3}-\left(-\frac{3}{10}\right)$

19. $-\frac{7}{9}+\frac{3}{4}-\frac{5}{18}$

20. $\frac{1}{3}-\frac{5}{8}-\left(-\frac{3}{4}\right)$

21. $1.92 - 6.8$

22. $-7.69 - (-5.823) - 13.513$

23. $-\frac{7}{12}-\left(-\frac{11}{24}\right)$

24. $\frac{5}{9}-\left(-\frac{5}{12}\right)+\frac{5}{18}$

1. ____________
2. ____________
3. ____________
4. ____________
5. ____________
6. ____________
7. ____________
8. ____________
9. ____________
10. ____________
11. ____________
12. ____________
13. ____________
14. ____________
15. ____________
16. ____________
17. ____________
18. ____________
19. ____________
20. ____________
21. ____________
22. ____________
23. ____________
24. ____________

Name Score

**Write as a fraction and a decimal.**

**1.** 30% **2.** 44% **3.** 110%

**4.** 140% **5.** 21% **6.** 15%

**1.** ____________
**2.** ____________
**3.** ____________
**4.** ____________
**5.** ____________
**6.** ____________

**Write as a fraction.**

**7.** $22\frac{2}{9}\%$ **8.** $4\frac{2}{5}\%$ **9.** $34\frac{1}{4}\%$

**7.** ____________
**8.** ____________
**9.** ____________

**Write as a decimal.**

**10.** 3.8% **11.** 16.17% **12.** 32.07

**10.** ____________
**11.** ____________
**12.** ____________

**Write as a percent.**

**13.** 1.18 **14.** 2.054 **15.** 0.009

**16.** $\frac{3}{5}$ **17.** $1\frac{1}{3}$ **18.** $3\frac{3}{4}$

**19.** $\frac{13}{25}$ **20.** $\frac{3}{14}$ **21.** $3\frac{1}{8}$

**13.** ____________
**14.** ____________
**15.** ____________
**16.** ____________
**17.** ____________
**18.** ____________
**19.** ____________
**20.** ____________
**21.** ____________

Name Score

**Evaluate.**

**1.** $4^2$

**2.** $5^3$

**3.** $-5^2$

**4.** $-3^4$

**5.** $(-2)^4$

**6.** $(-3)^3$

**7.** $(-5)^2$

**8.** $(-4)^2$

**9.** $-(-2)^3$

**10.** $(-3)^3$

**11.** $(12)^3$

**12.** $(-2)^2 \cdot 4^3$

**13.** $(-1)^3 \cdot 2^2$

**14.** $(-1)^2 \cdot 2^3$

**15.** $(-2)^4 \cdot 5^2$

**16.** $(-5) \cdot 4^3$

**17.** $2^2 \cdot 3^2 \cdot (-4)^2$

**18.** $(-3) \cdot 2^4$

**19.** $(-3) \cdot (-2)^2$

**20.** $(-3) \cdot (-2)^3$

**21.** $2^2 \cdot 3^2 \cdot (-5)$

**22.** $(-2)^2 \cdot 3^2 \cdot 6$

**23.** $(-6) \cdot 2^3 \cdot 3^2$

**24.** $(-2) \cdot 3^2 \cdot (-2)^2$

**1.** __________
**2.** __________
**3.** __________
**4.** __________
**5.** __________
**6.** __________
**7.** __________
**8.** __________
**9.** __________
**10.** __________
**11.** __________
**12.** __________
**13.** __________
**14.** __________
**15.** __________
**16.** __________
**17.** __________
**18.** __________
**19.** __________
**20.** __________
**21.** __________
**22.** __________
**23.** __________
**24.** __________

Name Score

**Simplify by using the Order of Operation Agreement.**

**1.** $9-3\times 2$

**2.** $3^2\cdot 2-4$

**3.** $3(5-1)-(-2)^2$

**4.** $15-16\div 2^2$

**5.** $21-15\div 3+4$

**6.** $10-(-2)^3-(-1)$

**7.** $12+18\div(-3)-1$

**8.** $10-2^3-(2-5)$

**9.** $5-3[9-(5-3)]$

**10.** $-3^2+2[8\div(1+3)]$

**11.** $20\div 5-3^2+(-2)^3$

**12.** $25\div(14-3^2)+(-5)$

**13.** $8(-9)+[3(7-3)^2]$

**14.** $11+\dfrac{19-3}{3^2-1}-7$

**15.** $32\div\dfrac{4^2}{7-5}-(-6)$

**16.** $48\div 4[11-(6-1)]-(-2)^2$

**17.** $3\cdot[15-(8-5)]\div 6$

**18.** $\dfrac{(-9)+(-7)}{5^2-17}\div(3-5)$

**19.** $-3^2+10[20\div(2-7)]$

**20.** $5\cdot[20-(9-4)]\div 25$

**21.** $6\cdot[14-(11-9)]\div 3^2$

**22.** $15+21\div(-7)-3$

**23.** $12-3^2-(2-7)$

**24.** $(2.54-2.09)^2\div 0.5+0.9$

1. __________
2. __________
3. __________
4. __________
5. __________
6. __________
7. __________
8. __________
9. __________
10. __________
11. __________
12. __________
13. __________
14. __________
15. __________
16. __________
17. __________
18. __________
19. __________
20. __________
21. __________
22. __________
23. __________
24. __________

Name Score

**Evaluate the variable expression when $a = -2$, $b = 3$, and $c = 4$.**

1. $-2a+3b$
2. $a^2-2$
3. $5(b-a)$
4. $4a+c^2$
5. $-2ab \div c$
6. $c^2-ab$
7. $(b+c)^2$
8. $(c-a)^2+2b$
9. $\frac{c-a}{b}$
10. $b^2-2ac$
11. $-\frac{3}{4}c^2+\frac{2}{3}b$
12. $\frac{a+2b}{c}$

1. ____________
2. ____________
3. ____________
4. ____________
5. ____________
6. ____________
7. ____________
8. ____________
9. ____________
10. ____________
11. ____________
12. ____________

**Evaluate the variable expression when $a = 3$, $b = -1$, $c = 2$ and $d = -4$.**

13. $2ad-4c$
14. $(c-b)^2-2d$
15. $2(b-d)+3a$
16. $\frac{d-b}{ac}$
17. $\frac{a-d}{b}$
18. $\frac{b^2-4c}{-2b-a^2}$
19. $\frac{1}{4}d^2+3ac$
20. $\frac{5}{9}a^2-bcd$
21. $\frac{-3cd}{ab^3}$
22. $c-\frac{1}{5}(ad-bc)$
23. $(d-a)^2+(b-c)^2$
24. $\frac{ab-cd}{a-bc}$

13. ____________
14. ____________
15. ____________
16. ____________
17. ____________
18. ____________
19. ____________
20. ____________
21. ____________
22. ____________
23. ____________
24. ____________

Name Score

**Use the given Property of the Real Numbers to complete the statement.**

**1.** The Associative Property of Addition
$(7+3)+5=?+(3+5)$

**2.** The Commutative Property of Multiplication
$8\cdot 5=5\cdot ?$

**3.** The Multiplication Property of One
$7\cdot ?=7$

**4.** The Inverse Property of Addition
$x+?=0$

**5.** The Multiplication Property of Zero
$6\cdot 0=?$

**6.** The Commutative Property of Addition
$xy+wz=?+xy$

**7.** The Associative Property of Multiplication
$4(5x)=?\cdot x$

**8.** The Inverse Property of Multiplication
$\frac{1}{x}(x)=?$

**9.** The Addition Property of Zero
$4+?=4$

**10.** The Distributive Property
$3(x+2)=?\cdot x+6$

**Identify the property that justifies the statement.**

**11.** $-11+11=0$

**12.** $d(f+g)=df+dg$

**13.** $(-45y)(0)=0$

**14.** $a\cdot 1=a$

**15.** $(a+b)+c=a+(b+c)$

**16.** $-7+0=-7$

**17.** $a(bc)=(ab)c$

**18.** $mn\left(\frac{1}{mn}\right)=1$

**19.** $a+b=b+a$

**20.** $(a\cdot b)\cdot c=c\cdot(a\cdot b)$

**1.** ______________
**2.** ______________
**3.** ______________
**4.** ______________
**5.** ______________
**6.** ______________
**7.** ______________
**8.** ______________
**9.** ______________
**10.** ______________
**11.** ______________
**12.** ______________
**13.** ______________
**14.** ______________
**15.** ______________
**16.** ______________
**17.** ______________
**18.** ______________
**19.** ______________
**20.** ______________

Name Score

**Simplify.**

**1.** $4x+7x$

**2.** $10x+11x$

**3.** $7a-2a$

**4.** $-2a+9a$

**5.** $7ab-5ab$

**6.** $-10xy+15xy$

**7.** $9xy+(-11xy)$

**8.** $-13xy+9xy$

**9.** $-8ab+13ab$

**10.** $-4mn+4mn$

**11.** $-\frac{1}{3}x-\frac{1}{4}x$

**12.** $-\frac{3}{5}x+\frac{2}{10}x$

**13.** $\frac{1}{4}x^2-\frac{7}{12}x^2$

**14.** $\frac{5}{8}x-\frac{1}{2}x$

**15.** $-\frac{2}{3}x^2-\frac{3}{4}x^2$

**16.** $6a-2a+4a$

**17.** $-3x^2-8x^2-2x^2$

**18.** $-2y^2+7y^2-6y^2$

**19.** $6x+(-5x)+2y$

**20.** $7y+(-8x)+3x$

**21.** $7y+7x-7y$

**22.** $2a+(-5b)-7b+4a$

**23.** $9y+(-11x)-6y+4y$

**24.** $x^2-8x+(-3x^2)+7x$

**25.** $-4b+3a-9b+11a$

**1.** ____________

**2.** ____________

**3.** ____________

**4.** ____________

**5.** ____________

**6.** ____________

**7.** ____________

**8.** ____________

**9.** ____________

**10.** ____________

**11.** ____________

**12.** ____________

**13.** ____________

**14.** ____________

**15.** ____________

**16.** ____________

**17.** ____________

**18.** ____________

**19.** ____________

**20.** ____________

**21.** ____________

**22.** ____________

**23.** ____________

**24.** ____________

**25.** ____________

Name Score

**Simplify.**

1. $2(5x)$
2. $6(4x)$
3. $-3(2a)$
4. $-3(-4y)$
5. $-7(-8y)$
6. $-6(2x^2)$
7. $-7(9x^2)$
8. $\frac{1}{5}(5x^2)$
9. $\frac{1}{4}(4x^2)$
10. $\frac{1}{9}(9a)$
11. $\frac{1}{10}(10x)$
12. $-\frac{1}{3}(-3x)$
13. $-\frac{1}{5}(-5a)$
14. $-\frac{1}{10}(-10a)$
15. $(7x)\left(\frac{1}{7}\right)$
16. $(15x)\left(\frac{1}{15}\right)$
17. $(-8y)\left(-\frac{1}{8}\right)$
18. $(-20n)\left(-\frac{1}{20}\right)$
19. $\frac{1}{5}(15x)$
20. $\frac{1}{6}(24x)$
21. $-\frac{1}{3}(12x)$
22. $-\frac{1}{6}(18x)$
23. $-\frac{7}{8}(32a^2)$
24. $-\frac{3}{5}(25x^2)$
25. $-\frac{1}{3}(-15y)$
26. $-\frac{4}{5}(-10y)$
27. $(52y)\left(\frac{1}{13}\right)$

1. ______
2. ______
3. ______
4. ______
5. ______
6. ______
7. ______
8. ______
9. ______
10. ______
11. ______
12. ______
13. ______
14. ______
15. ______
16. ______
17. ______
18. ______
19. ______
20. ______
21. ______
22. ______
23. ______
24. ______
25. ______
26. ______
27. ______

Name Score

**Simplify.**

**1.** $-(x+5)$

**2.** $3(2x-1)$

**3.** $4(2x-3)$

**4.** $-3(a+2)$

**5.** $-4(a+12)$

**6.** $-2(2y-5)$

**7.** $-4(3y-4)$

**8.** $3(-2x^2-12)$

**9.** $4(-5x^2-2)$

**10.** $-5(2y^2-3)$

**11.** $-7(3y^2-10)$

**12.** $-6(x^2+y^2)$

**13.** $-5(x^2-y^2)$

**14.** $-3(x^2-2y^2)$

**15.** $-2(5a^2-8b^2)$

**16.** $6(x^2+x-7)$

**17.** $3(x^2-2x-5)$

**18.** $-2(y^2+3y-5)$

**19.** $3(-a^2-a-4)$

**20.** $-7(-2x^2+4x+6)$

**21.** $-2(-3x^2+4x-1)$

**22.** $2(x^2+2xy-4y^2)$

**23.** $3(2x^2-3xy-4y^2)$

**24.** $-(2a^2+4a-3)$

**25.** $-(9b^2-5b+8)$

**1.** ______
**2.** ______
**3.** ______
**4.** ______
**5.** ______
**6.** ______
**7.** ______
**8.** ______
**9.** ______
**10.** ______
**11.** ______
**12.** ______
**13.** ______
**14.** ______
**15.** ______
**16.** ______
**17.** ______
**18.** ______
**19.** ______
**20.** ______
**21.** ______
**22.** ______
**23.** ______
**24.** ______
**25.** ______

Name Score

**Simplify.**

**1.** $5x-3(2x+7)$

**2.** $3a-(4a+1)$

**3.** $9-(10x-4)$

**4.** $7-2(3x-5)$

**5.** $6-(8+5y)$

**6.** $4n-(6-3n)$

**7.** $3x-(10-x)$

**8.** $2(x+1)-4(x-6)$

**9.** $3(x-3)-2(x+4)$

**10.** $8(y-1)+2(5-2y)$

**11.** $5(2y-5)-2(4-y)$

**12.** $2(x+y)-3(x-y)$

**13.** $3(a+b)-(a-2b)$

**14.** $3[x-3(x-2)]$

**15.** $5[x+3(x+6)]$

**16.** $-3[2x+3(2-x)]$

**17.** $-4[x+2(6-x)]$

**18.** $-2[3x-(x+8)]$

**19.** $-5[2x-(4x-1)]$

**20.** $3x-2[x-3(5-x)]$

**21.** $-6x+4[x-6(2-x)]$

**22.** $-4x-3[3x-3(x+6)]-5$

**23.** $3a-2[3b-(2b-a)]+4b$

**24.** $3a-3[b-(2b-a)]+5b$

**25.** $3x+2(x-2y)+4(2x-5y)$

**26.** $4y-2(y-2x)+3(6x-y)$

1. ______________
2. ______________
3. ______________
4. ______________
5. ______________
6. ______________
7. ______________
8. ______________
9. ______________
10. ______________
11. ______________
12. ______________
13. ______________
14. ______________
15. ______________
16. ______________
17. ______________
18. ______________
19. ______________
20. ______________
21. ______________
22. ______________
23. ______________
24. ______________
25. ______________
26. ______________

Name | Score

**Translate into a variable expression.**

1. the sum of 6 and $x$

2. $y$ less 10

3. $x$ less than 12

4. $t$ increased by 6

5. $a$ decreased by 1

6. 5 subtracted from $y$

7. $z$ divided by 8

8. $m$ multiplied by 11

9. 10 more than the square of $x$

10. 15 less than the cube of $x$

11. 5 times the sum of $n$ and 8

12. the sum of two-thirds of $x$ and 6

13. $x$ increased by the product of 3 and $x$

14. the quotient of –6 and $y$

15. the product of –2 and $t$

16. the product of 2 and the sum of $x$ and 8

17. the product of 5 and the total of $y$ and 6

18. 10 divided by the sum of $y$ and 5

19. 12 more than one-fourth of the square of $y$

20. 16 more than the product of $m$ and –1

21. $x$ increased by the quotient of $x$ and 2

22. three-fifths of the product of $w$ and 10

23. $y$ decreased by the product of $y$ and 2

24. the product of 7 and the total of $x$ and 3

1. ______
2. ______
3. ______
4. ______
5. ______
6. ______
7. ______
8. ______
9. ______
10. ______
11. ______
12. ______
13. ______
14. ______
15. ______
16. ______
17. ______
18. ______
19. ______
20. ______
21. ______
22. ______
23. ______
24. ______

Name Score

**Translate into a variable expression. Then simplify.**

1. a number subtracted from the product of four and the number

2. a number added to the product of five and the number

3. three less than the sum of a number and twelve

4. a number increased by the difference between eleven and the number

5. a number plus the sum of the number and seven

6. a number minus the sum of the number and five

7. the sum of two-fifths of a number and three-tenths of the number

8. the difference between three-eighths of a number and one-fourth of the number

9. three less than the total of a number and nine

10. four more than the total of a number and three

11. four times the sum of five times a number and six

12. the difference between eleven times a number and seven times the number

13. twelve more than the sum of ten and a number

14. fifteen decreased by the sum of a number and five

15. three times the sum of two consecutive integers

16. nine more than the sum of two consecutive integers

17. one-half of the sum of two consecutive even integers

18. twenty more than a number added to the difference between the number and nine

19. a number subtracted from the product of the number and five

20. six times the sum of the square of a number and four

1. ______
2. ______
3. ______
4. ______
5. ______
6. ______
7. ______
8. ______
9. ______
10. ______
11. ______
12. ______
13. ______
14. ______
15. ______
16. ______
17. ______
18. ______
19. ______
20. ______

Name Score

**Solve.**

1. Ten gallons of paint were poured into two containers of different sizes. Use one variable to express the amount poured into each container.

2. A wire 35 feet long was cut into two pieces of different lengths. Use one variable to express the lengths of the two pieces.

3. The speed of a commercial jet is three times the speed of a corporate jet. Express the speed of the commercial jet in terms of the speed of the corporate jet.

4. A banker divided $4200 between two accounts, one paying 4% annual interest and the second paying 3.5% annual interest. Express the amount invested in the 4% account in terms of the amount invested in the 3.5% account.

5. The base of a triangle is 4 feet longer than the height of the triangle. Express the base of the triangle in terms of the height of the triangle.

6. The length of a rectangular piece of poster board is 4 times the width. Express the length of the paper in terms of the width.

7. A board 8 feet long was cut into two pieces. Express the length of the longer piece in terms of the length of the shorter piece.

8. The length of a rectangular area rug is 3 feet less than twice the width. Express the length of the rug in terms of the width of the rug.

9. The number of cherry trees in an orchard is approximately one-fifth the number of apple trees in the orchard. Express the number of cherry trees in the orchard in terms of the number of apple trees in the orchard.

10. The number of adjunct faculty at a community college is 206 more than the number of full-time faculty. Express the number of adjunct faculty in terms of the number of full-time faculty at the community college.

1. ____________
2. ____________
3. ____________
4. ____________
5. ____________
6. ____________
7. ____________
8. ____________
9. ____________
10. ____________

Name Score

**Solve.**

1. Is 2 a solution of $3a = 6$?

2. Is 5 a solution of $x - 3 = 2$?

3. Is –2 a solution of $x - 3 = -1$?

4. Is –3 a solution of $2a + 8 = 2$?

5. Is 4 a solution of $3x + 4 = 4x$?

6. Is –1 a solution of $2a + 3 = 3a + 4$?

7. Is 0 a solution of $5 - m = 3m + 3$?

8. Is 3 a solution of $3 - 2n = n + 12$?

9. Is –2 a solution of $4 - 2m = 6 - m$?

10. Is 1 a solution of $5a + 3 = 2 + 4a$?

11. Is 2 a solution of $x^2 = x + 2$?

12. Is –3 a solution of $y^2 - 1 = y + 11$?

13. Is 6 a solution of $x^2 + 3 = 9 + 5x$?

14. Is 4 a solution of $2x^2 - 1 = 4x + 15$?

15. Is $\frac{3}{4}$ a solution of $3x - 1 = 4 - x$?

16. Is $-\frac{1}{4}$ a solution of $4t + 1 = 8t + 2$?

17. Is 3 a solution of $x^2 + 4x + 4 = (x + 2)^2$?

18. Is –5 a solution of $(x - 3)^2 = x^2 + 6x + 9$?

19. Is 3 a solution of $x(x - 1) = x^2 - 3$?

20. Is $-\frac{1}{2}$ a solution of $6t + 1 = -2$?

21. Is $\frac{2}{5}$ a solution of $10m + 5 = 5m + 3$?

22. Is $\frac{1}{3}$ a solution of $2x + 1 = 2 - x$?

23. Is $\frac{1}{4}$ a solution $n + 2 = 5n + 1$?

24. Is $-\frac{1}{2}$ a solution of $2x(x - 1) = x - 1$?

1. ______
2. ______
3. ______
4. ______
5. ______
6. ______
7. ______
8. ______
9. ______
10. ______
11. ______
12. ______
13. ______
14. ______
15. ______
16. ______
17. ______
18. ______
19. ______
20. ______
21. ______
22. ______
23. ______
24. ______

Name Score

**Solve and check.**

1. $x+3=8$
2. $x+4=7$
3. $a-4=11$
4. $y-5=9$
5. $1+a=10$
6. $m+7=2$
7. $t+8=0$
8. $n-3=-1$
9. $x-7=-4$
10. $x-9=-5$
11. $x-5=-3$
12. $x+4=4$
13. $a-2=-7$
14. $x-4=-2$
15. $z+7=1$
16. $t-5=-3$
17. $9+a=15$
18. $-6=n+2$
19. $3=m-7$
20. $5=-10+b$
21. $-9=-3+x$
22. $12=-4+a$
23. $c+\frac{4}{5}=-\frac{1}{5}$
24. $x-\frac{1}{8}=\frac{1}{8}$
25. $x-\frac{1}{5}=\frac{2}{5}$
26. $b+\frac{1}{2}=-\frac{1}{3}$
27. $x+\frac{1}{3}=-\frac{5}{6}$

1. ______
2. ______
3. ______
4. ______
5. ______
6. ______
7. ______
8. ______
9. ______
10. ______
11. ______
12. ______
13. ______
14. ______
15. ______
16. ______
17. ______
18. ______
19. ______
20. ______
21. ______
22. ______
23. ______
24. ______
25. ______
26. ______
27. ______

Name Score

**Solve and check.**

**1.** $4x = 12$ **2.** $5y = 30$ **3.** $2a = -10$

**4.** $3a = -15$ **5.** $-6m = 18$ **6.** $-2x = -24$

**7.** $-5n = -35$ **8.** $-48 = -8y$ **9.** $-42 = 6a$

**10.** $-18 = -6y$ **11.** $-\frac{y}{3} = 2$ **12.** $-\frac{b}{4} = 1$

**13.** $\frac{2}{3}y = 8$ **14.** $\frac{3}{5}x = 12$ **15.** $-\frac{3}{4}d = 9$

**16.** $-\frac{4}{5}m = 12$ **17.** $\frac{2x}{7} = 4$ **18.** $-\frac{2z}{5} = 8$

**19.** $-\frac{7z}{8} = 14$ **20.** $\frac{3n}{4} = 6$ **21.** $\frac{5x}{6} = -5$

**22.** $\frac{-3z}{5} = 18$ **23.** $-9 = -\frac{3x}{4}$ **24.** $-12 = \frac{-3a}{4}$

**25.** $\frac{2x}{3} = 1$ **26.** $6d - 3d = 6$ **27.** $9x - 5x = 20$

**1.** ______
**2.** ______
**3.** ______
**4.** ______
**5.** ______
**6.** ______
**7.** ______
**8.** ______
**9.** ______
**10.** ______
**11.** ______
**12.** ______
**13.** ______
**14.** ______
**15.** ______
**16.** ______
**17.** ______
**18.** ______
**19.** ______
**20.** ______
**21.** ______
**22.** ______
**23.** ______
**24.** ______
**25.** ______
**26.** ______
**27.** ______

Name Score

**Solve.**

**1.** 4 is what percent of 25?

**2.** What percent of 75 is 15?

**3.** Find 16% of 30.

**4.** What is 30% of 90?

**5.** 15% of what is 60?

**6.** 25% of what is 8?

**7.** What is 12.5% of 48?

**8.** 20 is what percent of 50?

**9.** 20% of what is 8?

**10.** 25% of what is 16?

**11.** 60 is what percent of 12?

**12.** 24 is what percent of 6?

**13.** Find 18% of 40.

**14.** What is 27% of 14?

**15.** Find 30.8% of 60.

**16.** A company built and equipped a new suite of offices for $200,000. Of this amount, $40,000 is for equipment and furnishings. What percent of the total expenditure was for equipment and furnishings?

**17.** A used car dealer estimates that 65% of the company's sales will occur on Thursday, Friday or Saturday. Using this estimate, how many cars out of the 80 sold last month were sold on Thursday, Friday or Saturday?

**18.** Approximately 70% of the employees in a company purchase supplementary dental insurance. Using this estimate, how many of the company's 200 employees purchased dental insurance?

**19.** A computer dealer estimates that 12.5% of the computers sold will require service during the one year warranty period. Using this estimate, how many computers were sold in a year in which 75 new computers were serviced?

**20.** The number of part-time employees in a business is 40. Last year the number of part-time employees was 50. What percent decrease does this represent?

**21.** The number of building permits issued in a city this year was 72,000. This represents 125% of last year's number of permits. How many building permits were issued last year?

**1.** ______
**2.** ______
**3.** ______
**4.** ______
**5.** ______
**6.** ______
**7.** ______
**8.** ______
**9.** ______
**10.** ______
**11.** ______
**12.** ______
**13.** ______
**14.** ______
**15.** ______
**16.** ______
**17.** ______
**18.** ______
**19.** ______
**20.** ______
**21.** ______

Name Score

**Solve and check.**

**1.** $2x+3=15$ **2.** $3y+5=26$ **3.** $4a-7=9$

**4.** $5m-16=19$ **5.** $3=4a+15$ **6.** $3x-7=-22$

**7.** $6n-9=-51$ **8.** $9=5+2d$ **9.** $17=9+4z$

**10.** $8-c=7$ **11.** $6-2w=-4$ **12.** $7-5x=-23$

**13.** $9-4t=1$ **14.** $16-7x=2$ **15.** $5y-35=0$

**16.** $12+3b=0$ **17.** $14+2m=0$ **18.** $-3x+7=-5$

**19.** $-4d+5=-31$ **20.** $-9x-4=-22$ **21.** $-11x+20=-2$

**22.** $-15=-12y+21$ **23.** $3=7-4a$ **24.** $2=12-5n$

**25.** $-33=-6b+3$ **26.** $-7x+2=-26$ **27.** $-4x-36=0$

**1.** ____________
**2.** ____________
**3.** ____________
**4.** ____________
**5.** ____________
**6.** ____________
**7.** ____________
**8.** ____________
**9.** ____________
**10.** ____________
**11.** ____________
**12.** ____________
**13.** ____________
**14.** ____________
**15.** ____________
**16.** ____________
**17.** ____________
**18.** ____________
**19.** ____________
**20.** ____________
**21.** ____________
**22.** ____________
**23.** ____________
**24.** ____________
**25.** ____________
**26.** ____________
**27.** ____________

Name Score

**Solve and check.**

1. $7x+4=3x+32$
2. $5y+1=y+17$
3. $10m+2=9m+10$
4. $4x-3=2x+7$
5. $8a-9=2a+15$
6. $11y-3=8y-9$
7. $12b-3=4b-27$
8. $14x-1=3x-23$
9. $6a-4=a-19$
10. $4x+3=13-x$
11. $3x-4=-22-3x$
12. $5y-1=-17-3y$
13. $2b+9=5b+15$
14. $m+3=3m+11$
15. $5d-3=7d+9$
16. $5y-7=2y-7$
17. $4a+11=a+11$
18. $5-4x=7-2x$
19. $9-5n=15-2n$
20. $4+7x=12+3x$
21. $3x-5=8x$
22. $2a-15=7a$
23. $7m=4m+24$
24. $8y=4y+40$
25. $-2x-3=3x+7$
26. $-6a-1=2a+23$
27. $-7n+4=-4n-14$

1. ______
2. ______
3. ______
4. ______
5. ______
6. ______
7. ______
8. ______
9. ______
10. ______
11. ______
12. ______
13. ______
14. ______
15. ______
16. ______
17. ______
18. ______
19. ______
20. ______
21. ______
22. ______
23. ______
24. ______
25. ______
26. ______
27. ______

Name Score

**Solve and check.**

**1.** $3x + 2(x - 1) = 8$ **2.** $5y + 3(y - 3) = 15$ **3.** $8n - 5(2n + 1) = 7$

**4.** $10x - 3(2x + 5) = 13$ **5.** $8m - 5(m - 2) - 7$ **6.** $5a - (2a - 7) = 10$

**7.** $7n - 3(4n - 1) = 13$ **8.** $3(2x - 1) - 2 = 13$ **9.** $2(3b + 2) - 7 = 9$

**10.** $4(2 - 3y) + 3y = 26$ **11.** $5(1 - 2x) + 5 = 80$ **12.** $9x + 1 = 3(2x + 7) - 2$

**13.** $4y - 5 = 6 + 3(y - 1)$ **14.** $11x + 4 = 2(3x - 2) - 7$ **15.** $8 - 7x = 14 - (5x + 6)$

**16.** $2x - 3 = 5(3x + 2)$ **17.** $4n - 9 = 3(2n + 5)$ **18.** $4b + 2(3b - 5) = 2b + 6$

**19.** $x + 2(3x - 1) = 6x - 5$ **20.** $3 - 2a = 10 - 3(2a - 3)$ **21.** $8 - 4x = 11 - (5x + 6)$

**22.** $4b + 3(2b - 1) = 2b + 13$ **23.** $x + 2(3x - 2) = 6x - 2$ **24.** $2y - 5 = 4(y - 7) + 5$

**25.** $3a - 4 = 3(2a + 1) + 5$ **26.** $4 - (8 - 5x) = 3x - 8$ **27.** $6 - (4 - 7x) = 3x + 2$

**1.** ______
**2.** ______
**3.** ______
**4.** ______
**5.** ______
**6.** ______
**7.** ______
**8.** ______
**9.** ______
**10.** ______
**11.** ______
**12.** ______
**13.** ______
**14.** ______
**15.** ______
**16.** ______
**17.** ______
**18.** ______
**19.** ______
**20.** ______
**21.** ______
**22.** ______
**23.** ______
**24.** ______
**25.** ______
**26.** ______
**27.** ______

Name Score

**Translate into an equation and solve.**

**1.** The sum of a number and nine is fifteen. Find the number.

**2.** Two more than a number is eleven. Find the number.

**3.** The difference between a number and six is ten. Find the number.

**4.** One-half of a number is nine. Find the number.

**5.** Five more than four times a number is thirteen. Find the number.

**6.** Two times the sum of a number and three is twelve. Find the number.

**7.** Eight less than three times a number is equal to seven. Find the number.

**8.** The quotient of a number and six is the opposite of two. Find the number.

**9.** The sum of seven times a number and three is the opposite of eighteen. Find the number.

**10.** The sum of five times a number and the number is twenty-four. Find the number.

**11.** The difference between six times a number and three times the number is twelve. Find the number.

**12.** Thirty is five minus the product of five and a number. Find the number.

**13.** Nine is one more than the product of two and a number. Find the number.

**14.** The sum of two numbers is twenty-one. Twice the larger is four times the smaller number. Find the two numbers.

**15.** Five times a number is equal to eighteen more than twice the number. Find the number.

**16.** Four times the sum of a number and three equals two less than five times the number. Find the number.

**1.** ____________

**2.** ____________

**3.** ____________

**4.** ____________

**5.** ____________

**6.** ____________

**7.** ____________

**8.** ____________

**9.** ____________

**10.** ____________

**11.** ____________

**12.** ____________

**13.** ____________

**14.** ____________

**15.** ____________

**16.** ____________

Name Score

**Solve.**

**1.** A total of \$8000 is deposited into two simple interest accounts. On one account, the annual interest rate is 10%, while on the second account the annual simple interest rate is 12%. How much should be invested in each account so that the total annual interest earned is \$900?

**2.** An investment club invested a part of \$14,000 in a 9% annual simple interest account and the remainder in a 15% annual simple interest account. The amount of interest earned for one year was \$1800. How much was invested in each account?

**3.** An investment of \$6000 is made at an annual simple interest rate of 14%. How much additional money must be invested at an annual simple interest rate of 10% so that the total interest earned is 12% of the total investment?

**4.** A total of \$12,000 is invested into two simple interest accounts. On one account the annual simple interest rate is 12%, while on the second account the annual interest rate is 18%. How much should be invested in each account so that the interest earned is the same?

**5.** A total of \$7000 is invested into two simple interest accounts. On one account the annual simple interest rate is 15%, while on the second account the annual simple interest rate is 20%. How much should be invested in each account so that the interest earned by each account is the same?

**6.** An accountant deposited an amount of money into a 10% annual simple interest account. Another deposit, \$4000 more than the first, was placed in a 16% annual interest account. The total interest earned on both investments for one year was \$2200. How much money was deposited into the 10% account?

**7.** An investment counselor for a corporation invested 70% of the company's investment account in 13% short term certificates. The remainder was invested in 12% corporate bonds. The annual interest earned from the two investments was \$261,000. What was the total amount invested?

**8.** An investment of \$20,000 is made in a 10.5% simple interest account. How much additional money must be deposited into an 8% simple interest account so that the total interest earned on both accounts is 9% of the total investment?

**1.** __________
**2.** __________
**3.** __________
**4.** __________
**5.** __________
**6.** __________
**7.** __________
**8.** __________

Name Score

**Solve.**

**1.** Two small planes start from the same point and fly in opposite directions. The first plane is 30 mph slower than the second plane. In three hours the planes are 1170 mi apart. Find the rate of each plane.

**2.** Two trains traveling on parallel tracks start at the same time from two towns 300 miles apart and travel toward each other. One of the trains travels 20 mph faster than the other. If they meet in 3 h, how fast is each traveling?

**1.** ______________

**2.** ______________

**3.** A family drove to a beach resort at an average speed of 44 mph and later returned over the same road at an average speed of 55 mph. Find the distance to the resort if the total driving time was 9 h.

**4.** Two cars, one traveling 12 mph faster than the other, start at the same time from the same point and travel in opposite directions. In four hours they are 384 mi apart. Find the rate of each car.

**3.** ______________

**4.** ______________

**5.** A motorboat leaves a harbor and travels at an average speed of 10 mph toward a small island. Three hours later a speed boat leaves the same harbor and travels at a rate of 18 mph toward the same island. In how many hours after the speed boat leaves will the speed boat be alongside the motorboat?

**6.** A pilot flew to Newark and back. The average speed going was 320 mph and the average speed returning was 480 mph. Find the distance to Newark if the total flying time was 5 h.

**5.** ______________

**6.** ______________

**7.** On a 300-mile trip, a car traveled at an average speed of 55 mph and then reduced the speed to 40 mph for the remainder of the trip. The trip took a total of 6 h. For how long did the car travel at each speed?

**8.** An executive flew in a helicopter to the airport to board a plane. The helicopter's flying speed was 120 mph and the airplane's flying speed was 600 mph. The entire trip was 2160 mi and took 4 h. How far did the executive fly in the helicopter?

**7.** ______________

**8.** ______________

**9.** Two cyclists start at the same time from opposite ends of a course which is 42 mi long. One cyclist is riding at 16 mph and the second is riding at 12 mph. How long after they begin will they meet?

**10.** Two planes start at the same time from the same point and fly in opposite directions. The first plane is flying 30 mph faster than the second plane. In four hours they are 2880 mi apart. Find the rate of each plane.

**9.** ______________

**10.** ______________

Name Score

**Solve.**

**1.** The perimeter of a rectangle is 144 ft. The length of the rectangle is three times the width. Find the length and width of the rectangle.

**2.** The perimeter of a triangle is 42 ft. The first side is 3 ft more than the second side. The third side is 1 ft less than twice the second side. Find the measure of each side.

**3.** The perimeter of a rectangle is 76 m. The length of the rectangle is 5 m more than twice the width. Find the length and width of the rectangle.

**4.** The perimeter of a triangle is 180 ft. One side is twice the second side. The third side is 20 ft more than the second side. Find the measure of each side.

**5.** In an isosceles triangle, two sides are equal. The third side is 75% of the length of one of the equal sides. Find the length of each side when the perimeter is 165 cm.

**6.** The perimeter of a triangle is 180 cm. The length of the first side is twice the length of the second side. The length of the third side is three times the length of the second side. Find the length of each side.

**7.** The perimeter of a rectangle is 118 m. The width of the rectangle is 7 m less than the length. Find the length and width of the rectangle.

**8.** The width of a rectangle is 60% of the length. The perimeter of the rectangle is 128 ft. Find the length and width of the rectangle.

**9.** In an isosceles triangle, two sides are equal. The third side is 7 ft less than one of the equal sides. The perimeter is 53 ft. Find the length of each side.

**10.** The perimeter of a rectangle is 20 m. The length of the rectangle is 2 m more than the width. Find the measure of the width.

**1.** ______________

**2.** ______________

**3.** ______________

**4.** ______________

**5.** ______________

**6.** ______________

**7.** ______________

**8.** ______________

**9.** ______________

**10.** ______________

Name Score

**Solve.**

**1.** Find the supplement of a 111° angle.

**2.** Find the complement of a 90° angle.

**3.** Find the complement of a 9° angle.

**4.** Find the supplement of a 79° angle.

**Find *x*.**

**5.**

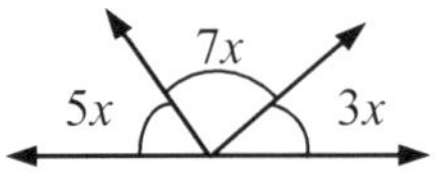


**6.**

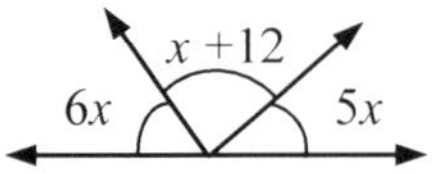


**7.**

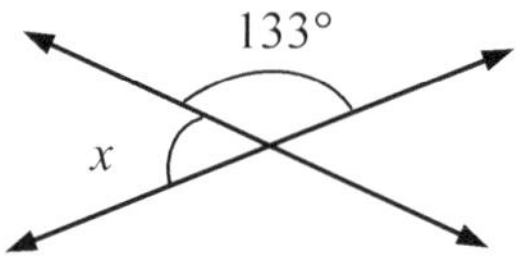


**8.**

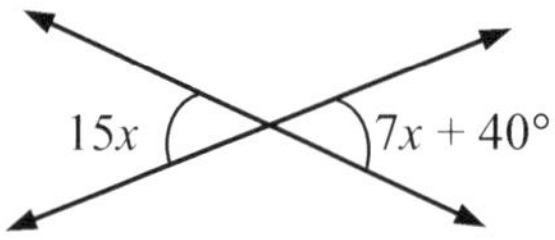


**Find the measures of angles *a* and *b*.** $k \parallel \ell$

**9.**

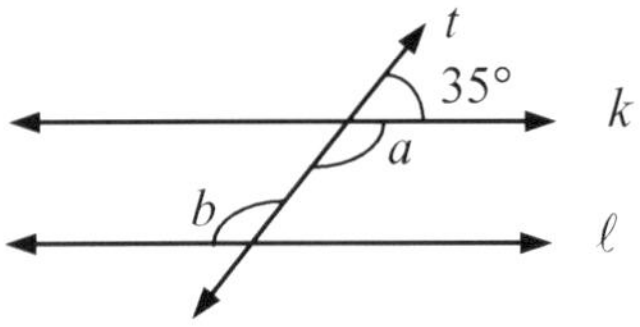


**10.**

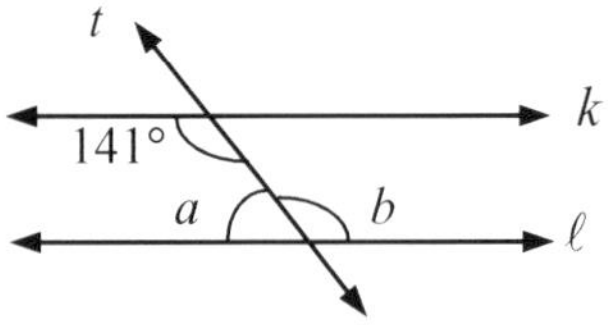


**Find *x*.** $k \parallel \ell$

**11.**

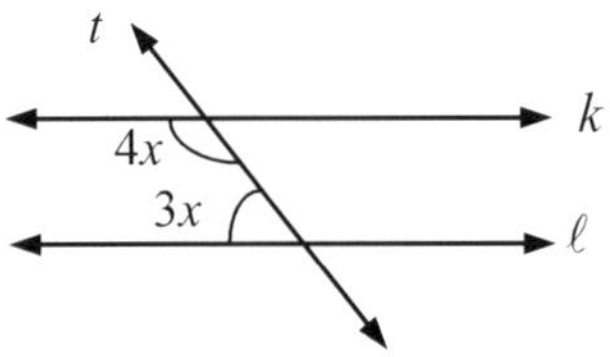


**12.**

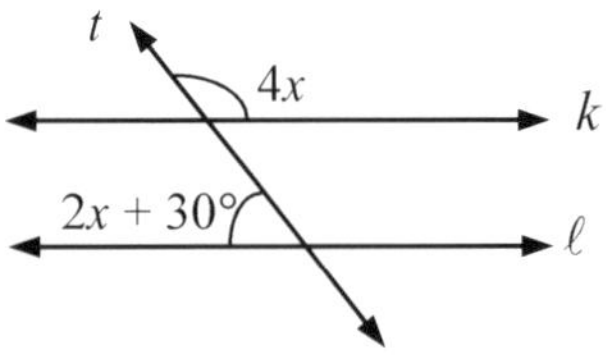


1. ____________
2. ____________
3. ____________
4. ____________
5. ____________
6. ____________
7. ____________
8. ____________
9. ____________
10. ____________
11. ____________
12. ____________

Name Score

**Solve.**

**1.** A triangle has a 40° angle and a right angle. What is the measure of the third angle?

**2.** A triangle has a 53° angle and a right angle. Find the measure of the third angle.

**3.** Two angles of a triangle measure 18° and 107°. Find the measure of the third angle.

**4.** Two angles of a triangle measure 68° and 43°. Find the measure of the third angle.

**5.** A triangle has a 121° angle and a 15° angle. What is the measure of the third angle?

**6.** A triangle has a 79° angle and a 72° angle. What is the measure of the third angle?

**7.** Two angles of a triangle measure 92° and 58°. What is the measure of the third angle?

**8.** A triangle has a 84° angle and a 39° angle. Find the measure of the third angle.

**1.** ______________

**2.** ______________

**3.** ______________

**4.** ______________

**5.** ______________

**6.** ______________

**7.** ______________

**8.** ______________

Name Score

**Solve.**

**1.** A grocery store buys a pound of coffee for $5.40 and uses a mark-up rate of 25%. Find the selling price.

**2.** A gift shop uses a mark-up rate of 48%. One lamp has a selling price of $37. Find the cost of the lamp.

**3.** The selling price of an action figure is $13. The mark-up rate used by the toy store is 30%. Find the cost of the action figure.

**4.** The manufacturer's cost for a digital camera is $180. The manufacturer then sells the digital camera for $252. What is the mark-up rate?

**5.** The manager of a produce stand uses a mark-up rate of 42%. What is the selling price of a bag of potatoes which cost $3?

**6.** A video game costs a retailer $24. Find the selling price when the mark-up rate is 35%.

**7.** A bakery sells a cake for $15.75. The bakery's cost to make the cake is $10.40. What is the mark-up rate?

**8.** The selling price for a jacket is $150. The cost to the clothing store for the jacket is $100. What is the mark-up rate?

**9.** The cost to a car dealer for a stereo radio is $120. The selling price of the radio is $160. What is the mark-up rate?

**10.** The cost to an appliance dealer for a coffee maker is $20. The dealer sells the coffee maker for $29. What is the mark-up rate?

**11.** The selling price for a set of golf clubs is $270. The clubs cost the golf pro $150. What is the mark-up rate?

**12.** The manager of an office furniture store uses a mark-up rate of 60%. What is the selling price of a file cabinet which costs $90?

**1.** ____________

**2.** ____________

**3.** ____________

**4.** ____________

**5.** ____________

**6.** ____________

**7.** ____________

**8.** ____________

**9.** ____________

**10.** ____________

**11.** ____________

**12.** ____________

Name Score

**Solve.**

**1.** A microwave oven which regularly sells for $200 is on sale for 20% off the regular price. Find the sale price.

**2.** A clock radio which regularly sells for $32 is on sale for 25% off the regular price. Find the sale price.

**3.** The sale price for a fire extinguisher is $28, which is 30% off the regular price. Find the regular price.

**4.** The sale price for a shredder is $56, which is 20% off the regular price. Find the regular price.

**5.** A bicycle which regularly sells for $120 is on sale for $96. Find the discount rate.

**6.** A cordless telephone which regularly sells for $80 is on sale for $60. Find the discount rate.

**7.** A pair of bowling shoes is on sale for $26. This is 35% off the regular price. Find the regular price.

**8.** During a sale, a clothing store reduced all items 40%. What was the regular price of a blazer which was on sale for $90?

**9.** A home computer which regularly sells for $400 is on sale for $340. Find the discount rate.

**10.** A bag of grass seed costs a retailer $12. Find the selling price when the mark-up rate is 45%.

**11.** A tennis racket is on sale for $24. This is $33\frac{1}{3}\%$ off the regular price. Find the regular price.

**12.** A lawn chair which regularly sells for $20 is on sale for $15.50. Find the discount rate.

**1.** ____________

**2.** ____________

**3.** ____________

**4.** ____________

**5.** ____________

**6.** ____________

**7.** ____________

**8.** ____________

**9.** ____________

**10.** ____________

**11.** ____________

**12.** ____________

Name Score

**Solve.**

**1.** A total of $8000 is deposited into two simple interest accounts. On one account, the annual interest rate is 10%, while on the second account the annual simple interest rate is 12%. How much should be invested in each account so that the total annual interest earned is $900?

**2.** An investment club invested a part of $14,000 in a 9% annual simple interest account and the remainder in a 15% annual simple interest account. The amount of interest earned for one year was $1800. How much was invested in each account?

**1.** ____________

**2.** ____________

**3.** An investment of $6000 is made at an annual simple interest rate of 14%. How much additional money must be invested at an annual simple interest rate of 10% so that the total interest earned is 12% of the total investment?

**4.** A total of $12,000 is invested into two simple interest accounts. On one account the annual simple interest rate is 12%, while on the second account the annual interest rate is 18%. How much should be invested in each account so that the interest earned is the same?

**3.** ____________

**4.** ____________

**5.** A total of $7000 is invested into two simple interest accounts. On one account the annual simple interest rate is 15%, while on the second account the annual simple interest rate is 20%. How much should be invested in each account so that the interest earned by each account is the same?

**6.** An accountant deposited an amount of money into a 10% annual simple interest account. Another deposit, $4000 more than the first, was placed in a 16% annual interest account. The total interest earned on both investments for one year was $2200. How much money was deposited into the 10% account?

**5.** ____________

**6.** ____________

**7.** An investment counselor for a corporation invested 70% of the company's investment account in 13% short term certificates. The remainder was invested in 12% corporate bonds. The annual interest earned from the two investments was $261,000. What was the total amount invested?

**8.** An investment of $20,000 is made in a 10.5% simple interest account. How much additional money must be deposited into an 8% simple interest account so that the total interest earned on both accounts is 9% of the total investment?

**7.** ____________

**8.** ____________

Name Score

**Solve.**

1. A chemist has some 12% hydrogen peroxide solution and some 9% hydrogen peroxide solution. How many milliliters of each should be mixed to make a 450 milliliter solution which is 10% hydrogen peroxide?

2. How many grams of pure acid must be added to 60 g of a 15% acid solution to make a solution which is 40% acid?

3. How many ounces of pure water must be added to 75 oz of a 20% salt solution to make a salt solution which is 15% salt?

4. A syrup manufacturer has some pure maple syrup and some which is 80% maple syrup. How many liters of each should be mixed to get 120 L which is 95% maple syrup?

5. A butcher has some burger which is 15% fat and some which is 25% fat. How may pounds of each should be mixed to make 60 lb of hamburger which is 18% fat?

6. A 110-pound bag of animal feed is 30% oats. How many pounds of oats must be added to this feed to produce a mixture which is 45% oats?

7. Forty-five ounces of pure grapefruit juice is added to 30 oz of a fruit punch which is 15% grapefruit juice. What is the percent concentration of grapefruit juice in the resulting mixture?

8. A silversmith mixes 60 g of one alloy which is 50% silver with 200 g of another silver alloy. The resulting alloy is 65% silver. Find the percent of silver in the 200 gram alloy.

9. A clothing manufacturer has some fiber which is 25% polyester and some which is 55% polyester. How many pounds of each fiber should be woven together to product 300 lb of a fabric which is 35% polyester?

10. A baker mixed some flour which was 30% wheat with 100 lb of flour which was 45% wheat to make a mixture which is 36% wheat. How many pounds of the 30% wheat flour were used?

1. ______________
2. ______________
3. ______________
4. ______________
5. ______________
6. ______________
7. ______________
8. ______________
9. ______________
10. ______________

Name Score

**Find the Additive Inverse.**

**1.** –3

**2.** –7

**3.** 5

**4.** 17

**5.** –87

**6.** 72

**7.** 66

**8.** 52

**9.** –120

**10.** –36

1. ____________
2. ____________
3. ____________
4. ____________
5. ____________
6. ____________
7. ____________
8. ____________
9. ____________
10. ____________

**Solve.**

**11.** Let $x \in \{-9, -3, 4\}$. For which values of $x$ is $x \geq -3$ true?

**12.** Let $y \in \{-8, -2, 1, 6\}$. For which values of $x$ is $y < 1$ true?

**13.** Let $a \in \{-60, 13, 28\}$. Evaluate $-a$ for each element of the set.

**14.** Let $a \in \{-48, -5, 9\}$. Evaluate $|\ b\ |$ for each element of the set.

11. ____________
12. ____________
13. ____________
14. ____________

Name Score

## Use the roster method to write the set.

**1.** the negative integer multiples of 6 that are greater than –23.

**2.** the positive integers less than 21 that are divisible by 4

**3.** the prime numbers between 20 and 30

1. ____________
2. ____________
3. ____________

## Use set builder notation to write the set.

**4.** the real numbers greater than or equal to –6

**5.** the real numbers between –1 and 1, inclusive

**6.** the integers between –10 and 10

4. ____________
5. ____________
6. ____________

## Find $A \cup B$.

**7.** $A = \{-6,\ 0,\ 6\}$, $B = \{-1,\ 0,\ 1\}$

**8.** $A = \{-3,\ -1,\ 0,\ 1\}$, $B = \{0,\ 1,\ 3\}$

**9.** $A = \{-2,\ -1,\ 0\}$, $B = \{-1,\ 0,\ 2\}$

7. ____________
8. ____________
9. ____________

## Find $A \cap B$.

**10.** $A = \{-6,\ 0,\ 6\}$, $B = \{-2,\ 0,\ 2\}$

**11.** $A = \{1,\ 2,\ 5,\ 10\}$, $B = \{5,\ 10,\ 50\}$

**12.** $A = \{1,\ 4,\ 9,\ 16\}$, $B = \{1,\ 4,\ 9\}$

10. ____________
11. ____________
12. ____________

## Graph the solution set.

**13.** $\{x \mid x \geq -3\}$

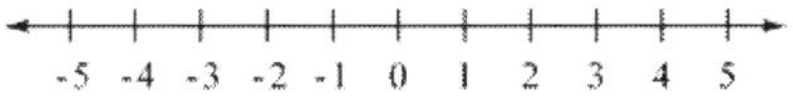


**14.** $\{x \mid x \geq -4\} \cap \{x \mid x \leq 3\}$

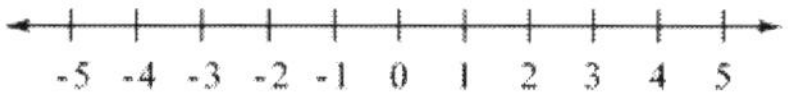


**15.** $\{x \mid x < 4\} \cap \{x \mid x > -1\}$

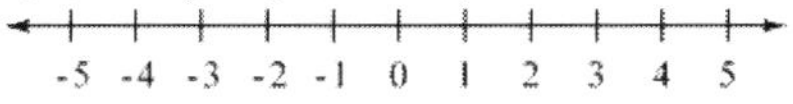


**16.** $\{x \mid x < -1\} \cup \{x \mid x > 1\}$

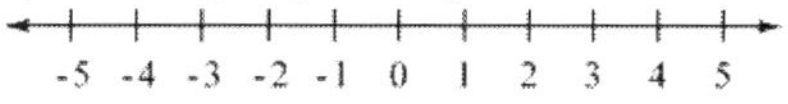


**17.** $\{x \mid x > -2\} \cap \{x \mid x \leq 3\}$

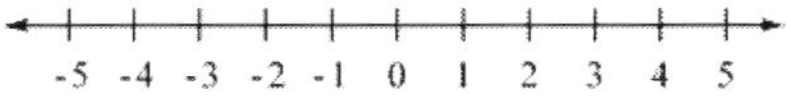


**18.** $\{x \mid x > 3\} \cup \{x \mid x > -3\}$

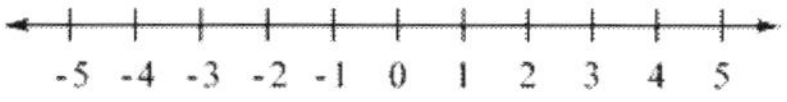


13. ____________
14. ____________
15. ____________
16. ____________
17. ____________
18. ____________

Name Score

**Solve.**

1. $x - 4 < 2$

2. $x + 5 \geq 2$

3. $2x \leq 10$

4. $3x > 15$

5. $-4x > 12$

6. $-5x \leq -20$

7. $7x - 1 > 2x + 9$

8. $6x + 1 \geq 4x - 3$

9. $3x - 2 > 7$

10. $2x + 3 < 15$

11. $4x - 3 \leq 9$

12. $5x + 1 \leq -9$

13. $7x + 2 > 4x - 7$

14. $6x + 5 < 4x - 5$

15. $5x + 7 > 4x - 2$

16. $9x + 6 < 2x - 8$

17. $7x + 3 \geq 3x + 19$

18. $6x - 1 < 2x + 7$

19. $7 - 6x \geq 31$

20. $2 - 7x \leq 23$

21. $-4 - 3x > -7$

22. $-3 - x < 5$

23. $7x - 3 < 5x - 11$

24. $5x + 4 \geq x - 16$

1. ______
2. ______
3. ______
4. ______
5. ______
6. ______
7. ______
8. ______
9. ______
10. ______
11. ______
12. ______
13. ______
14. ______
15. ______
16. ______
17. ______
18. ______
19. ______
20. ______
21. ______
22. ______
23. ______
24. ______

Name Score

**Solve. Write the solution in set builder notation.**

1. $2x < 8$ and $x + 3 > 1$
2. $x - 3 \le 5$ and $3x > -15$
3. $x + 1 \ge 7$ or $3x < 9$
4. $4x < 16$ or $x - 2 > 7$
5. $-2x > -12$ and $-4x < 12$
6. $\frac{1}{3}x > -1$ and $4x < 12$
7. $x + 3 \ge 6$ and $2x \ge 8$
8. $2x < -10$ and $x - 1 < 3$
9. $-3x > 9$ and $x + 2 > 5$
10. $3x + 1 < 7$ and $3x + 5 > -1$
11. $6x + 5 < 11$ or $3x - 1 > 8$
12. $5x - 3 < -18$ or $6x - 1 > 17$

1. ______
2. ______
3. ______
4. ______
5. ______
6. ______
7. ______
8. ______
9. ______
10. ______
11. ______
12. ______

**Solve. Write the solution in interval notation.**

13. $-2 < 3x + 7 < 10$
14. $-5 < 2x + 1 < 9$
15. $0 < 2x - 8 < 2$
16. $-2 < 3x + 7 < 4$
17. $3x - 2 > 7$ or $3x - 4 \le -10$
18. $5x - 7 > 9$ or $4x - 7 < -11$
19. $2x - 5 \ge 3$ and $3x - 2 > 10$
20. $5x - 2 < 3$ or $6x - 5 < 7$
21. $-7 \le 6x + 17 \le 35$
22. $2 \le 5x - 13 \le 12$
23. $5 - 8x \le 21$ and $4 - 5x > -1$
24. $7 - x \ge 9$ and $7 - 2x < 5$

13. ______
14. ______
15. ______
16. ______
17. ______
18. ______
19. ______
20. ______
21. ______
22. ______
23. ______
24. ______

Name Score

**Solve.**

**1.** The length of a rectangle is 3 ft more than five times the width. Express as an integer the maximum width of the rectangle when the perimeter is less than 54 ft.

**2.** Three times the difference between a number and five is less than or equal to eight times the sum of the number and 10. Find the smallest number that will satisfy the inequality.

**1.** __________

**2.** __________

**3.** Suppose Lemon Rental Cars rents sedans for \$39 per day with unlimited mileage, and Klunker Rentals offers sedans for \$24.99 per day but charges \$0.14 for each mile beyond 150 miles driven per day. You want to rent a car for a 10-day vacation. How many miles can you drive during the vacation if Klunker Rentals is to be less expensive than Lemon?

**4.** The temperature range for a week was between 23°F and 68°F. Find the temperature range in Celsius degrees.

$$C = \frac{5(F-32)}{9}$$

**3.** __________

**4.** __________

**5.** Heather Malen earns \$1500 per month plus 8% commission on the amount of sales. Heather's goal is to earn a minimum of \$4060 per month. What amount of sales will enable Heather to earn \$4060 or more per month?

**6.** Stronge National Bank offers two different checking accounts. Plan A charges \$2.50 per month and \$0.25 per check after the first 10 checks. Plan B charges \$7.50 per month with unlimited check writing. How many checks can be written per month if plan A is less expensive than plan B?

**5.** __________

**6.** __________

**7.** An average of 80 to 89 in an algebra class receives a B grade. A student has scores of 86, 95, 87, 79 on four tests. Find the range of scores on the fifth test that will give the student a B for the course.

**8.** Find three consecutive odd integers whose sum is between 78 and 99.

**7.** __________

**8.** __________

Name Score

**Solve.**

1. $|x| = 6$
2. $|a| = 3$
3. $|b| = 5$
4. $|c| = 11$
5. $|-y| = 7$
6. $|-t| = 2$
7. $|-a| = 9$
8. $|-x| = 1$
9. $|x| = -3$
10. $|-x| = -5$
11. $|5x - 10| = 0$
12. $|2x - 6| = 8$
13. $|4 - 2x| = 8$
14. $|a - 6| = 0$
15. $|3x - 2| = 0$
16. $|3 - 2x| = 5$
17. $|2 - 3x| = 7$
18. $|2a + 7| = 1$
19. $|5x - 1| - 1 = 3$
20. $|4x - 3| + 2 = 5$
21. $|3x + 2| - 1 = 3$
22. $|4x + 3| - 1 = 8$
23. $|5 - 3x| + 4 = 1$
24. $3 - |3x + 1| = 3$
25. $4 + |2x - 1| = 7$
26. $7 - |3x - 2| = 4$
27. $3 - |2x - 5| = 3$

1. ____________
2. ____________
3. ____________
4. ____________
5. ____________
6. ____________
7. ____________
8. ____________
9. ____________
10. ____________
11. ____________
12. ____________
13. ____________
14. ____________
15. ____________
16. ____________
17. ____________
18. ____________
19. ____________
20. ____________
21. ____________
22. ____________
23. ____________
24. ____________
25. ____________
26. ____________
27. ____________

Name Score

**Solve.**

**1.** $|x| > 2$

**2.** $|x| < 4$

**3.** $|3 - x| \geq 1$

**4.** $|2 - x| \geq 5$

**5.** $|2x - 1| < 7$

**6.** $|x + 7| \geq 3$

**7.** $|x - 3| < 2$

**8.** $|x - 4| \leq 7$

**9.** $|3x + 5| > 11$

**10.** $|5x - 1| > 14$

**11.** $|2x - 3| \leq -3$

**12.** $|4 - 3x| \leq 1$

**13.** $|3 - 4x| > 11$

**14.** $|3x - 2| < 10$

**15.** $|18 - 2x| \leq 0$

**16.** $|5x - 3| < 12$

**17.** $|3 - 2x| > 7$

**18.** $|4x - 1| < 3$

**1.** __________

**2.** __________

**3.** __________

**4.** __________

**5.** __________

**6.** __________

**7.** __________

**8.** __________

**9.** __________

**10.** __________

**11.** __________

**12.** __________

**13.** __________

**14.** __________

**15.** __________

**16.** __________

**17.** __________

**18.** __________

Name Score

**Solve.**

**1.** A doctor has prescribed 4 cc of medicine for a patient. The tolerance is 0.06 cc. Find the lower and upper limits of the amount of medication to be given.

**2.** The power of an electrical system is 105 volts plus or minus 20 volts. Find the lower and upper limits of voltage on which the system will run.

**3.** The power of a hand-held dryer is 115 volts plus or minus 17.5 volts. Find the lower and upper limits of voltage on which the dryer will run.

**4.** A doctor has prescribed 2.5 cc of medicine for a patient. The tolerance is 0.03 cc. Find the lower and upper limits of the amount of medicine to be given.

**5.** Find the lower and upper limits of a 28,000-ohm resistor with a 2% tolerance.

**6.** Find the lower and upper limits of a 16,000-ohm resistor with an 8% tolerance.

**7.** Find the lower and upper limits of a 72-ohm resistor with a 5% tolerance.

**8.** A small lamp is 6.25 volts plus or minus 0.5 volts. Find the lower and upper limits of voltage on which it will work.

**1.** ______________

**2.** ______________

**3.** ______________

**4.** ______________

**5.** ______________

**6.** ______________

**7.** ______________

**8.** ______________

Name Score

**Find the distance to the nearest hundredth between the given points.**

**1.** $P_1(1, 4)$ and $P_2(6, 2)$

**2.** $P_1(3, -2)$ and $P_2(-1, 1)$

**3.** $P_1(-2, 5)$ and $P_2(-1, -1)$

**4.** $P_1(-4, -1)$ and $P_2(-2, 3)$

**1.** ____________

**2.** ____________

**3.** ____________

**4.** ____________

**Find the coordinates of the midpoint of the line segment connecting the points.**

**5.** $P_1(-5, -3)$ and $P_2(3, 3)$

**6.** $P_1(4, 2)$ and $P_2(1, 1)$

**7.** $P_1(2, 5)$ and $P_2(-2, 1)$

**8.** $P_1(2, -3)$ and $P_2(-1, 4)$

**9.** $P_1(-1, 3)$ and $P_2(3, -1)$

**10.** $P_1(2, -2)$ and $P_2(1, -1)$

**5.** ____________

**6.** ____________

**7.** ____________

**8.** ____________

**9.** ____________

**10.** ____________

Name Score

**Solve.**

**1.** Graph the ordered pairs (2, 1) and (1, 3). Draw a line between the two points.

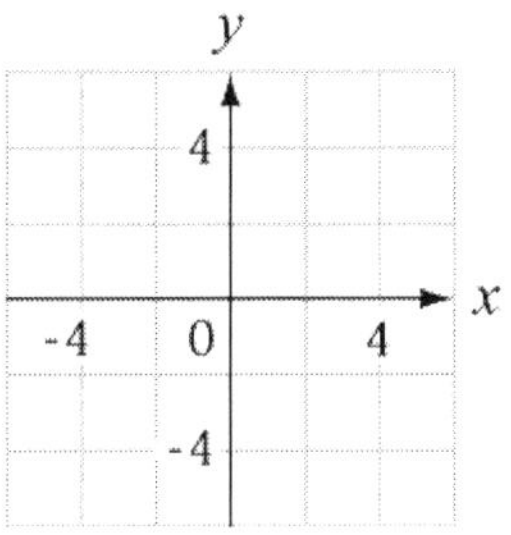


**2.** Graph the ordered pairs (–1, 2) and (2, –1). Draw a line between the two points.

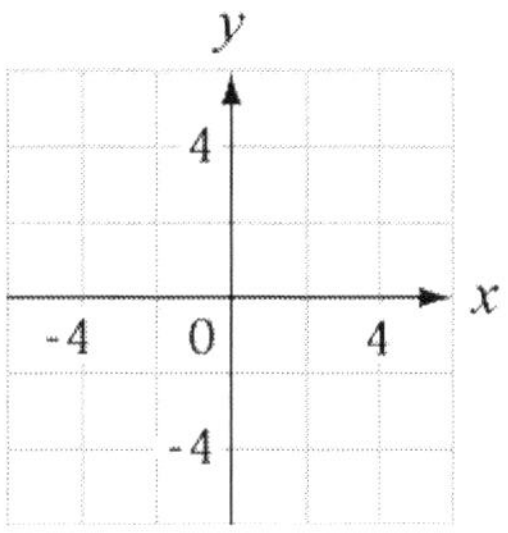


**3.** Graph the ordered pairs (–2, –3) and (3, 1). Draw a line between the two points.

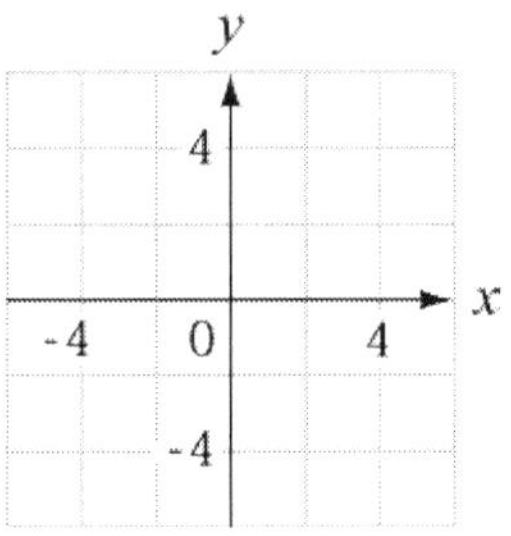


**4.** Graph the ordered pairs (0, –3) and (–2, 0). Draw a line between the two points.

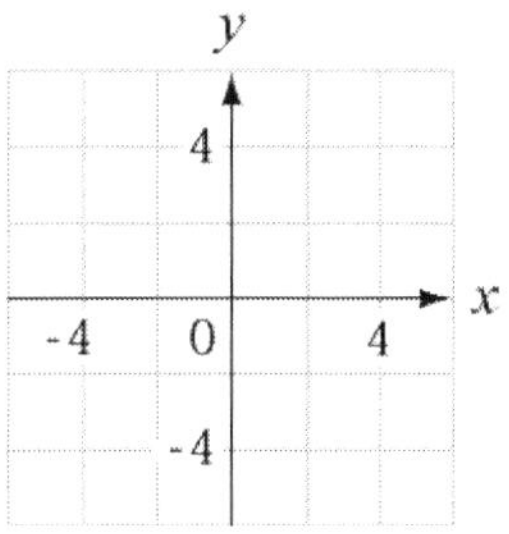


**5.** Find the coordinates of each of the points.

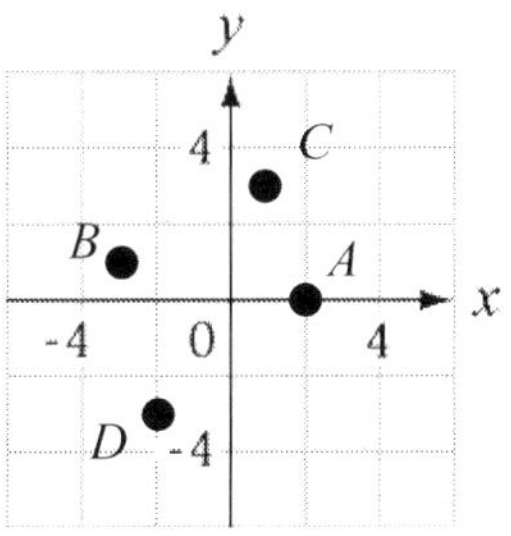


**6.** Find the coordinates of each of the points.

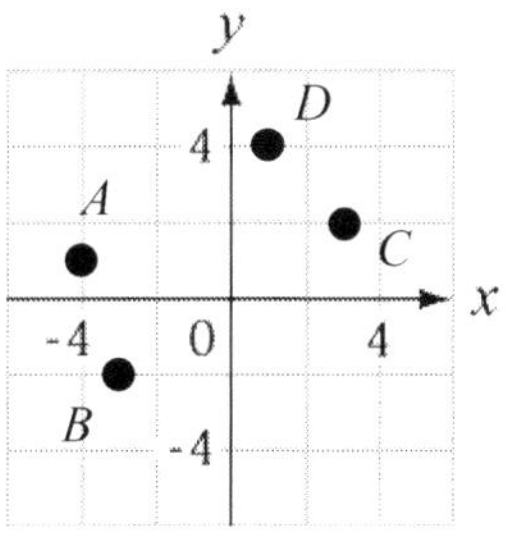


**5.** ______________

**6.** ______________

**7.** Draw a line through all points with an abscissa of –2.

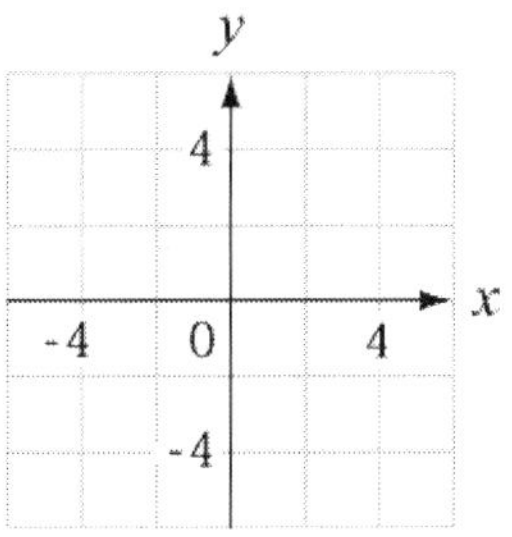


**8.** Draw a line through all points with an ordinate of 3.

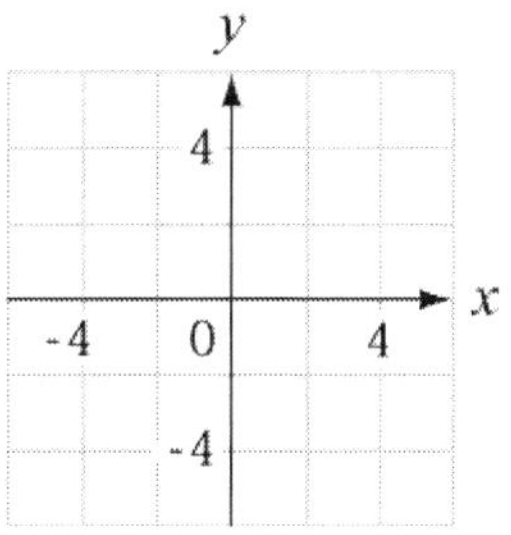

Name Score

**State whether the relation is a function.**

1. {(1, 1), (2, 3), (4, 3), (5, 7)}
2. {(–3, –1), (–1, –2), (1, 5), (2, 6), (1, –7)}

1. ____________
2. ____________

**Given $f(x) = 2x + 1$, evaluate:**

3. $f(-1)$
4. $f(2)$
5. $f(0)$

3. ____________
4. ____________
5. ____________

**Given $g(x) = x^2 + x - 1$, evaluate:**

6. $g(0)$
7. $g(1)$
8. $g(-1)$
9. $g(2)$
10. $g(-2)$
11. $g(t)$

6. ____________
7. ____________
8. ____________
9. ____________
10. ____________
11. ____________

**Find the domain and range of the function.**

12. {(1, 1), (2, 3), (4, 3), (5, 7)}
13. {(1, 4), (6, 2), (3, 1), (5, 4)}

12. ____________
13. ____________

**Find the range of the function defined by the equation and the given domain.**

14. $f(x) = 4x + 3$
    domain = {–1, 1, 3}
15. $g(x) = \frac{3}{4 - x}$
    domain = {–1, 0, 3}
16. $r(t) = t^2 - 3t + 1$
    domain = {–2, 1, 4}
17. $c(n) = \frac{3}{2}n - 1$
    domain = {–2, 0, 4}

14. ____________
15. ____________
16. ____________
17. ____________

Name Score

**Graph each function by evaluating the function at the given values of *x*. Plot the resulting ordered pairs. Then connect the points to form the graph.**

**1.** $f(x) = 2x - 4$,
$x = 0, 1, 2, 3, 4$

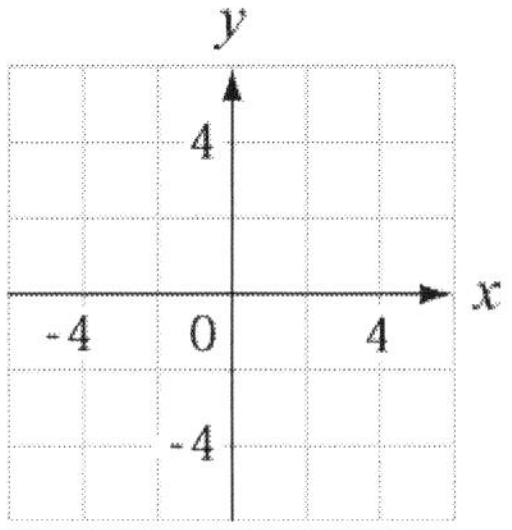


**2.** $g(x) = -3x - 3$
$x = -2, -1, 0$

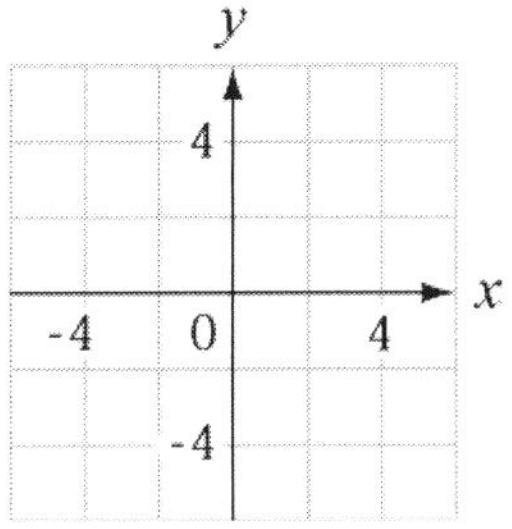


**3.** $h(x) = -x^2 + 3$
$x = -2, -1, 0, 1, 2$

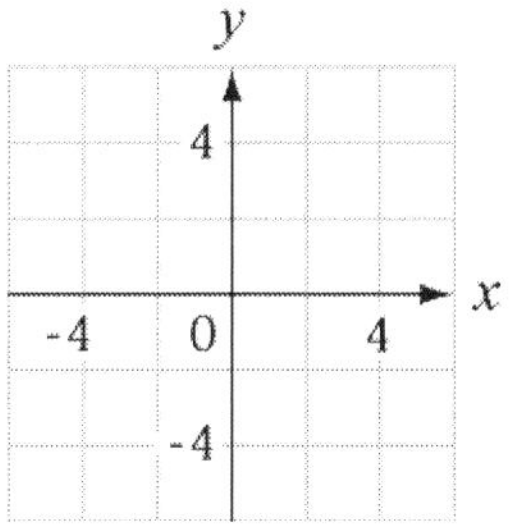


**4.** $P(x) = x^2 + 2x + 1$
$x = -3, -2, -1, 0, 1$

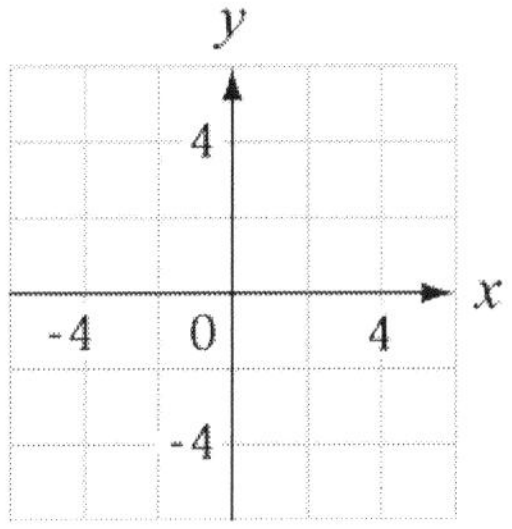


**5.** $f(x) = -|x - 1| - 2$
$x = -3, -2, -1, 0, 1, 2, 3, 4$

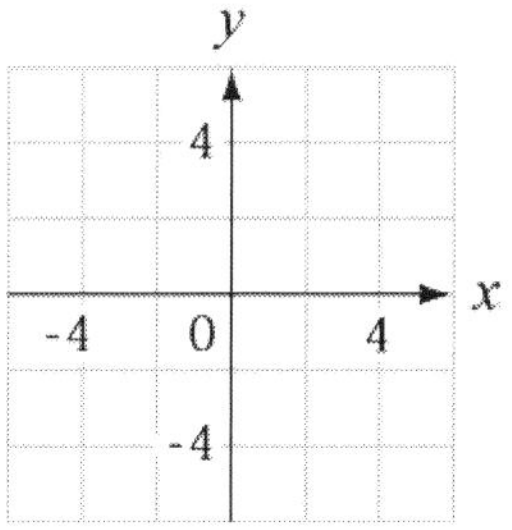


**6.** $g(x) = |x| + 2$
$x = -3, -2, -1, 0, 1, 2, 3$

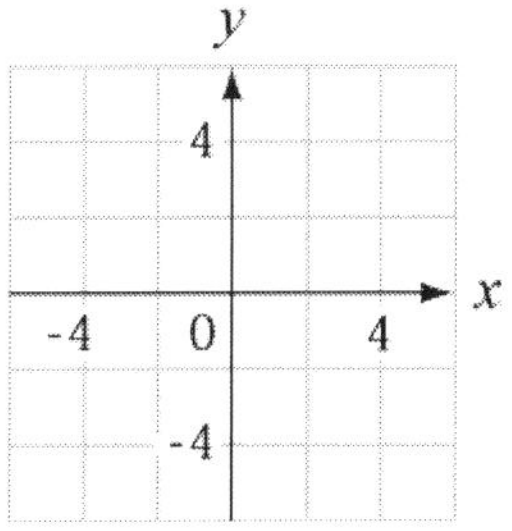

Name Score

**Use the Vertical Line Test to see if the graph represents function.**

**1.**

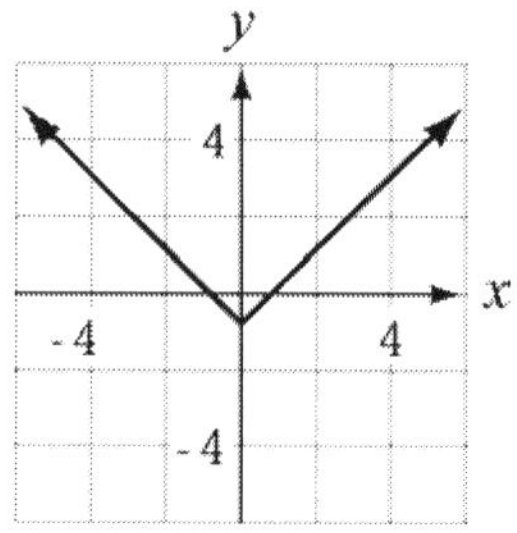


**2.**

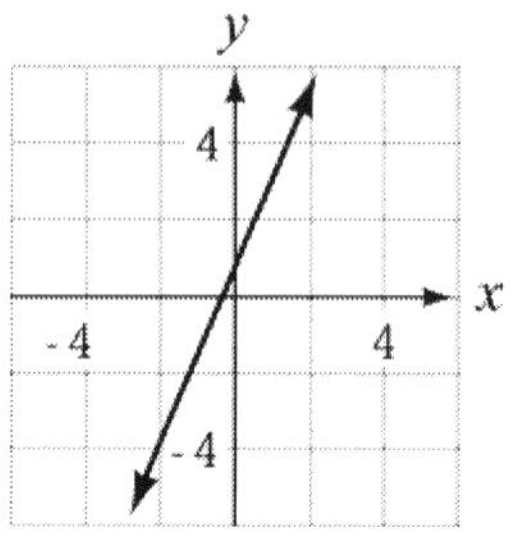


1. ______________

2. ______________

**3.**

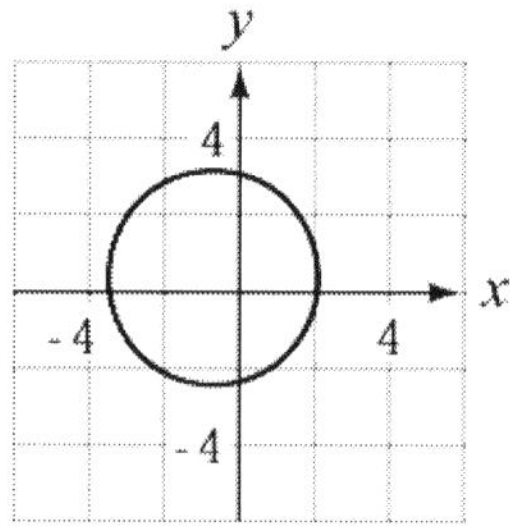


**4.**

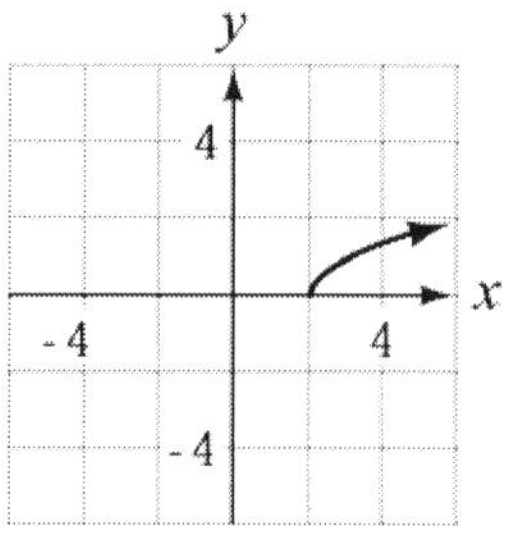


3. ______________

4. ______________

**5.**

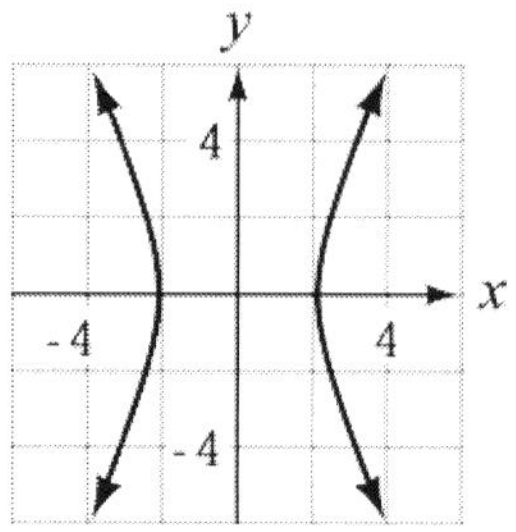


**6.**

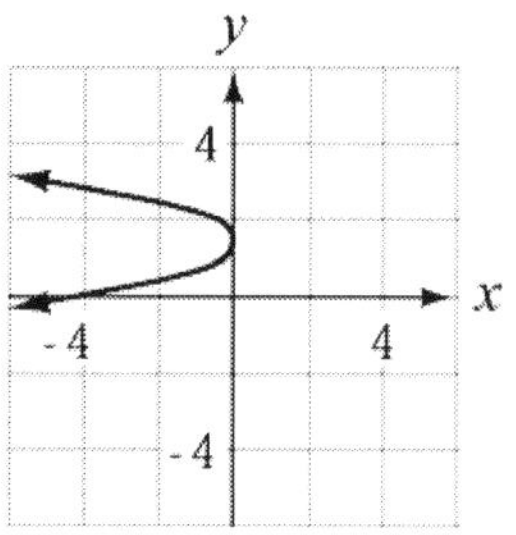


5. ______________

6. ______________

**7.**

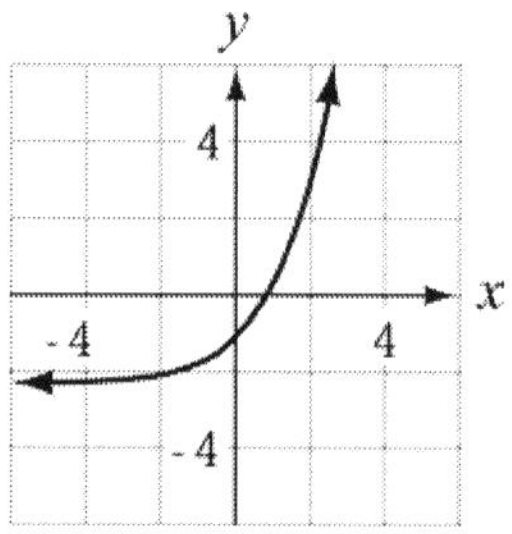


**8.**

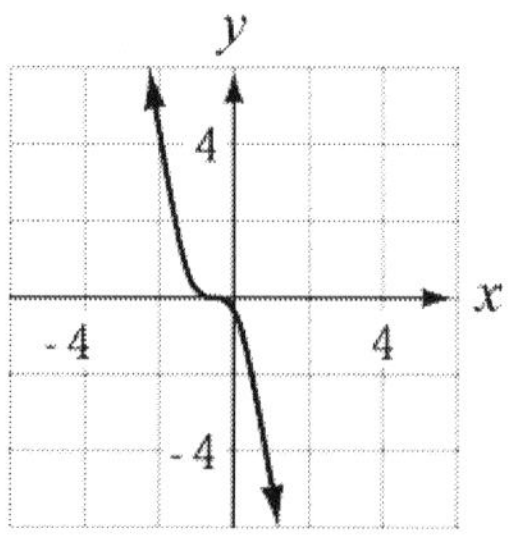


7. ______________

8. ______________

Name Score

**Graph.**

**1.** $y = 2x - 1$

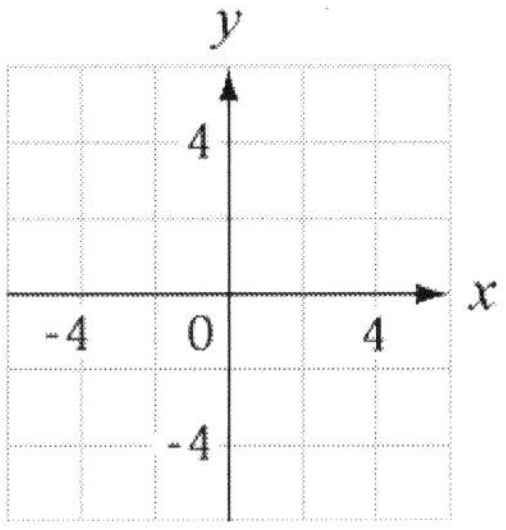


**2.** $y = 3x - 3$

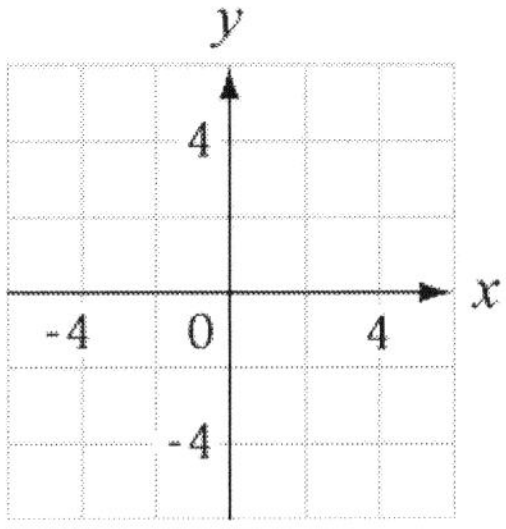


**3.** $y = \frac{1}{3}x$

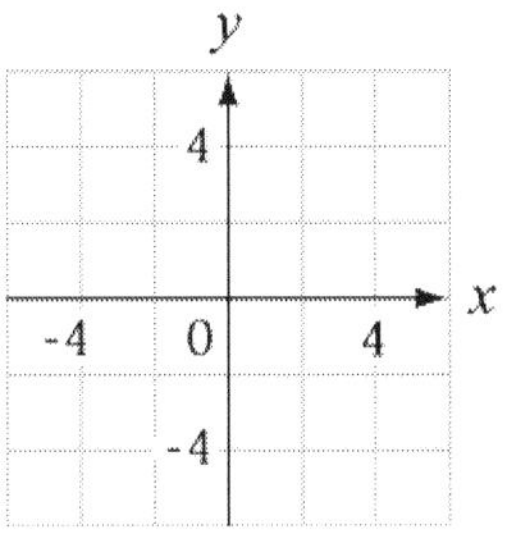


**4.** $y = \frac{4}{3}x$

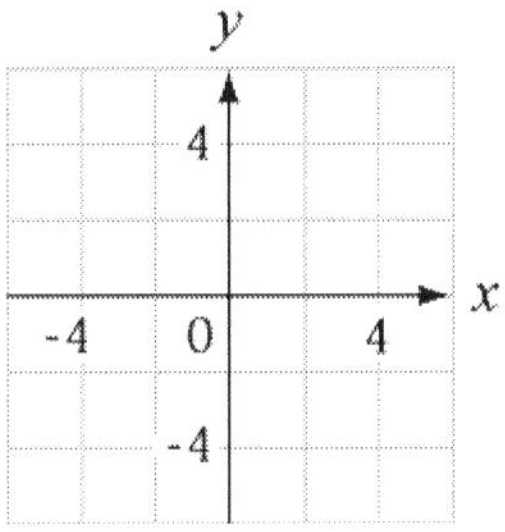


**5.** $y = -\frac{1}{3}x + 2$

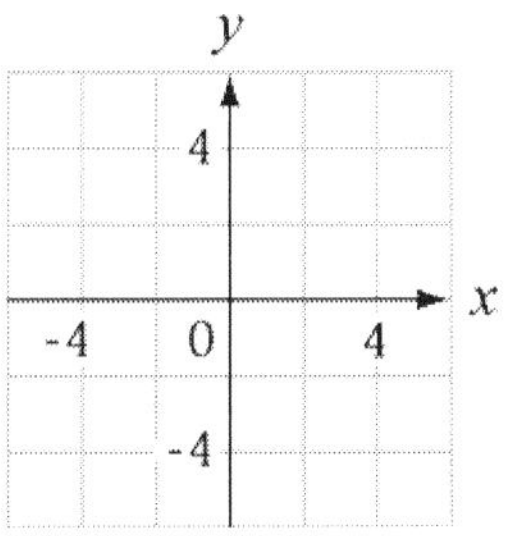


**6.** $y = \frac{1}{4}x - 1$

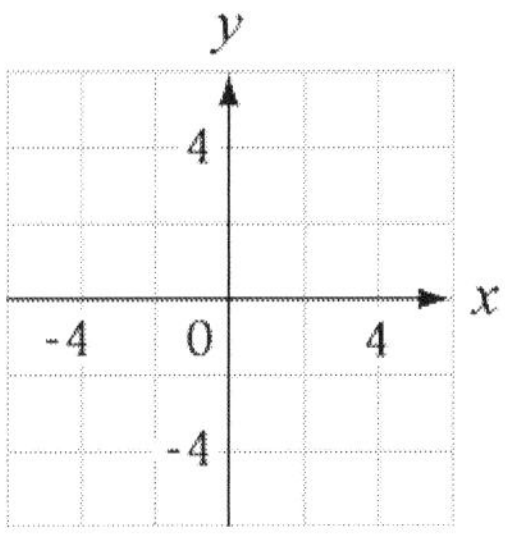


**7.** $y = -\frac{1}{2}x + 3$

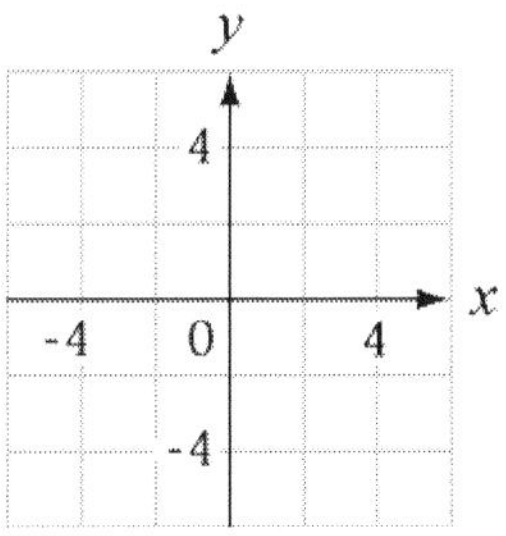


**8.** $y = -\frac{3}{2}x + 1$

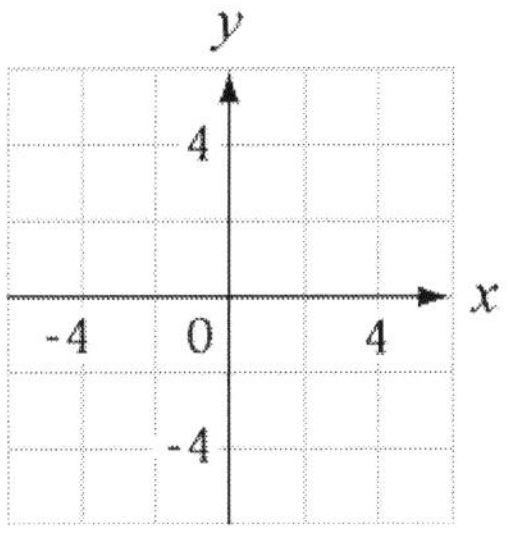

Name Score

**Graph.**

1. $3x - 2y = 6$

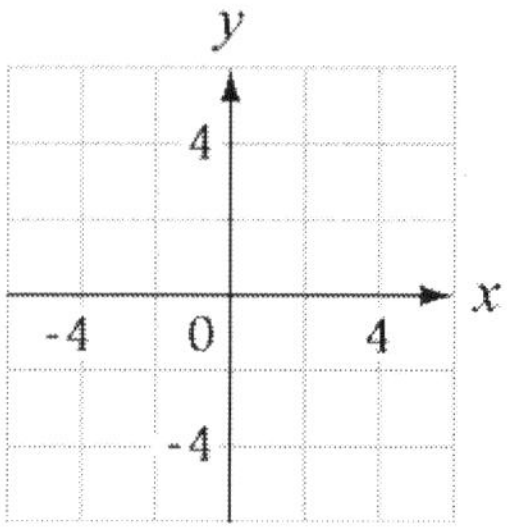


2. $y = 2$

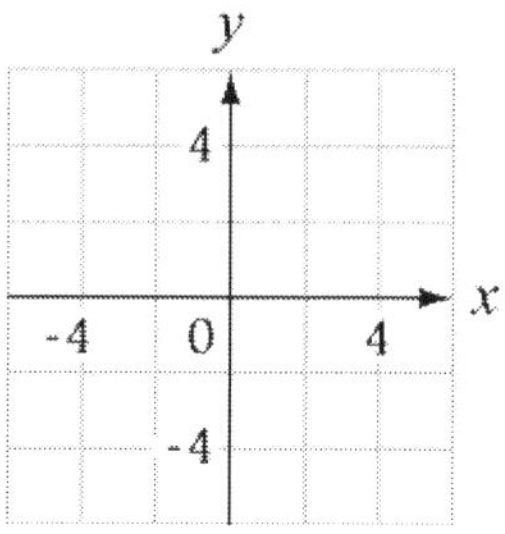


**Find the *x*– and *y*–intercepts and graph.**

3. $y = \frac{2}{5}x - 2$

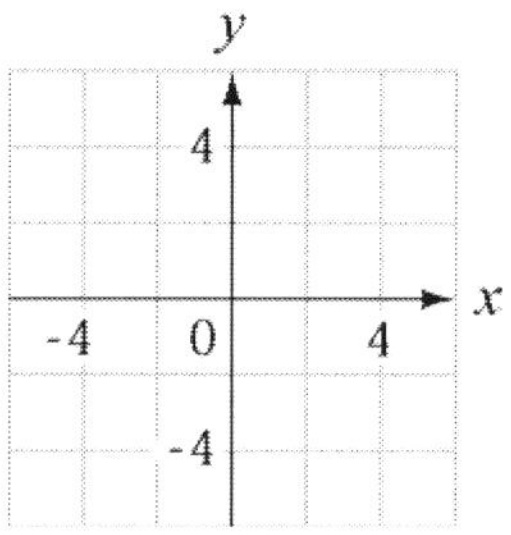


4. $3x - y = 6$

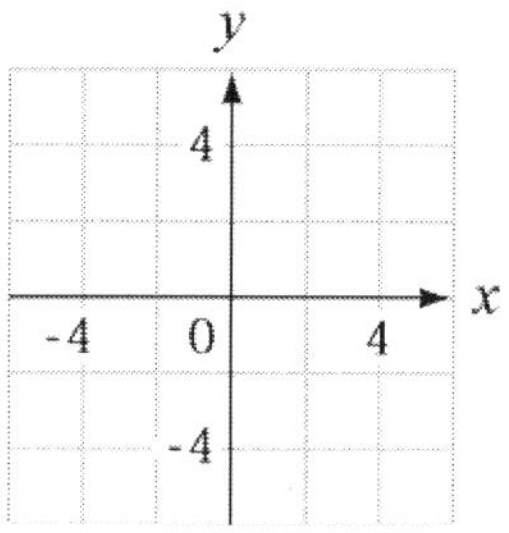


5. $y = 4x$

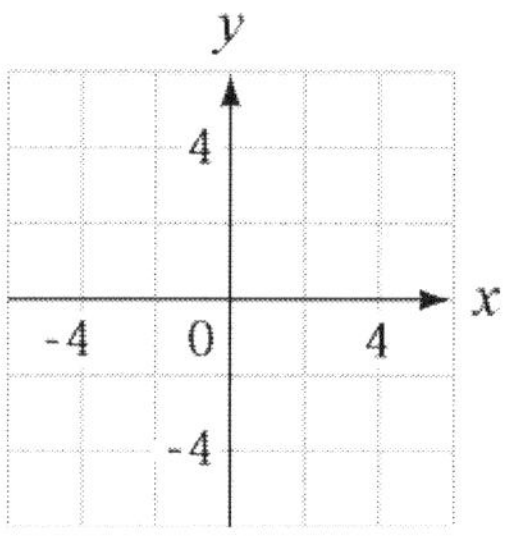


6. $2x - y = 5$

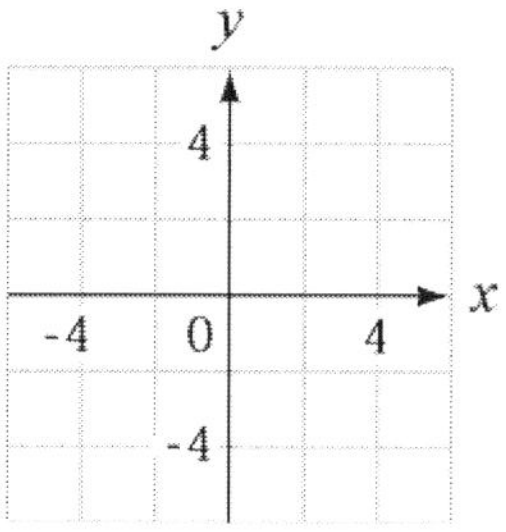


**Find the zero of each linear function.**

7. $f(x) = -4x + 12$

8. $g(x) = 2x - 7$

9. $r(t) = \frac{4}{3}t - 2$

10. $h(t) = -\frac{2}{5}t - 5$

7. ____________

8. ____________

9. ____________

10. ____________

Name Score

**Solve.**

**1.** Loren receives \$10 per hour as a mathematics tutor. The equation that describes her wages is $w = 10t$, where $t$ is the number of hours she spends tutoring. Graph this equation for $0 \le t \le 20$. The ordered pair (16, 160) is on the graph. Write a sentence that describes the meaning of this ordered pair.

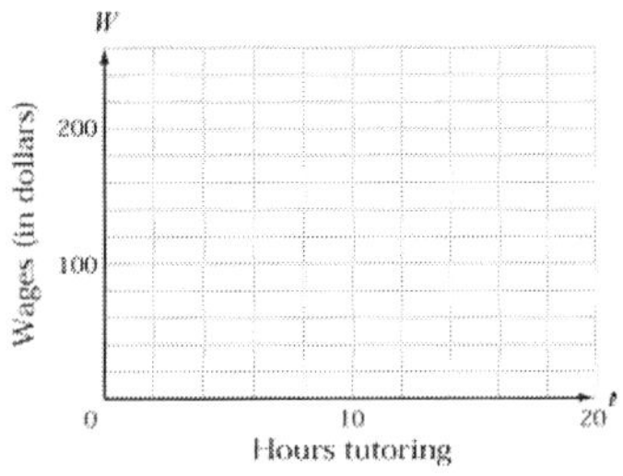


**1.** ____________

**2.** A roller coaster has a maximum speed of 99 ft/s. The equation that describes the total number of feet traveled by the roller coaster in $t$ seconds at this speed is given by $D = 99t$. Graph this equation for $0 \le t \le 10$. The point (6, 594) is on this graph. Write a sentence that describes the meaning of this ordered pair.

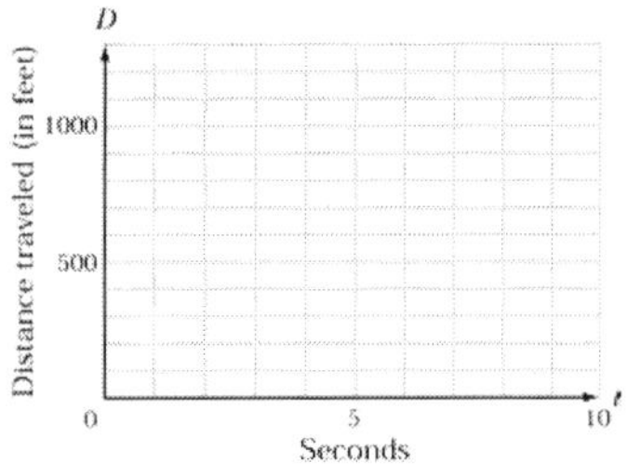


**2.** ____________

**3.** The monthly cost for receiving messages from a telephone answering service is \$7.00 plus \$0.25 per message. The equation that describes the cost is $C = 0.25n + 7.00$, where $n$ is the number of messages received. Graph the equation for $0 \le n \le 40$. The point (36, 16) is on the graph. Write a sentence that describes the meaning of this ordered pair.

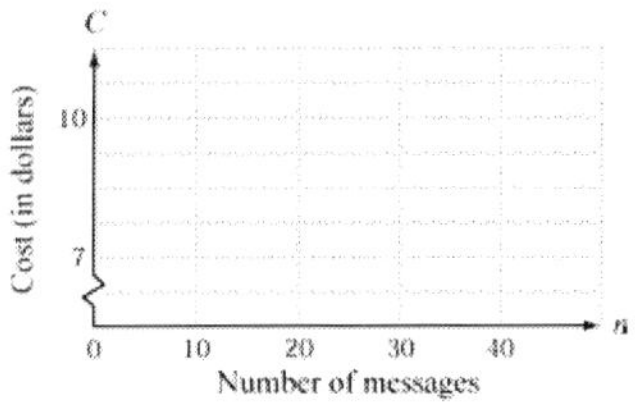


**3.** ____________

**4.** The cost of manufacturing snow boards is \$4500 for startup and \$75 per board manufactured. The equation that describes the cost of manufacturing $n$ boards is $C = 75n + 4500$. Graph this equation for $0 \le n \le 100$. The point (60, 9000) is on the graph. Write a sentence that describes the meaning of this ordered pair.

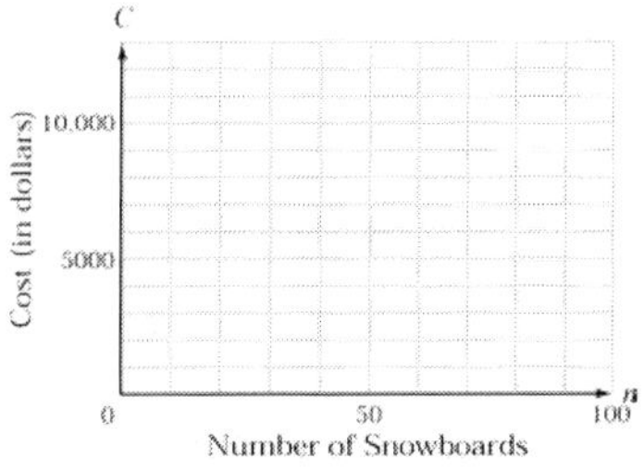


**4.** ____________

Name Score

**Find the slope of the line containing the points.**

**1.** $P_1(1, -3), P_2(5, -1)$

**2.** $P_1(-6, 7), P_2(2, 3)$

**3.** $P_1(0, -3), P_2(-1, 0)$

**4.** $P_1(1, 2), P_2(3, 4)$

**5.** $P_1(3, 2), P_2(2, 1)$

**6.** $P_1(-2, 1), P_2(-6, 3)$

**7.** $P_1(2, 3), P_2(3, 2)$

**8.** $P_1(4, -5), P_2(1, -4)$

**9.** $P_1(-1, -6), P_2(-4, -1)$

**10.** $P_1(-4, -5), P_2(-2, 0)$

**11.** $P_1(4, 3), P_2(2, 1)$

**12.** $P_1(1, 2), P_2(4, 8)$

**13.** $P_1(-4, 4), P_2(2, -2)$

**14.** $P_1(-3, 4), P_2(-1, -6)$

**15.** $P_1(0, -4), P_2(3, -10)$

**16.** $P_1(-1, 3), P_2(2, 4)$

**17.** $P_1(2, -2), P_2(0, 4)$

**18.** $P_1(-1, 2), P_2(-3, 4)$

**19.** $P_1(1, 2), P_2(-1, 2)$

**20.** $P_1(1, -2), P_2(5, -1)$

**21.** $P_1(2, 0), P_2(1, -1)$

**1.** ______
**2.** ______
**3.** ______
**4.** ______
**5.** ______
**6.** ______
**7.** ______
**8.** ______
**9.** ______
**10.** ______
**11.** ______
**12.** ______
**13.** ______
**14.** ______
**15.** ______
**16.** ______
**17.** ______
**18.** ______
**19.** ______
**20.** ______
**21.** ______

Name Score

**Graph by using the slope and the *y*–intercept.**

**1.** $y = \frac{1}{2}x + 3$

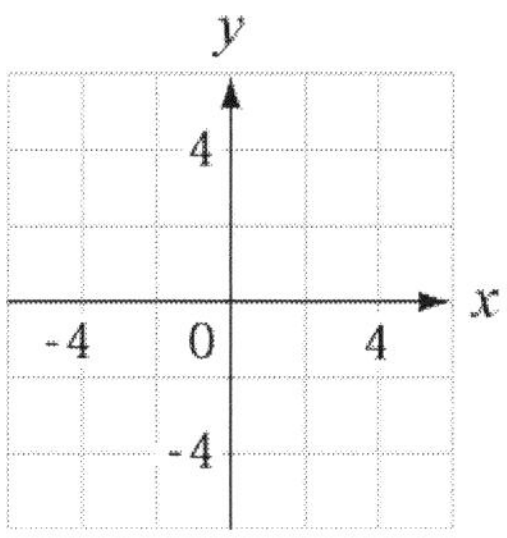


**2.** $y = \frac{3}{2}x + 2$

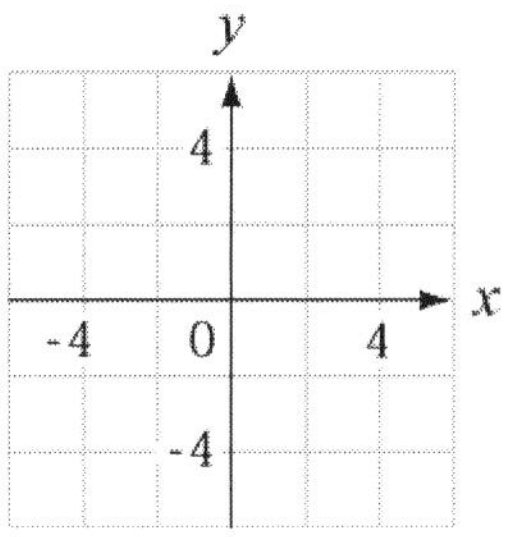


**3.** $y = -\frac{4}{3}x$

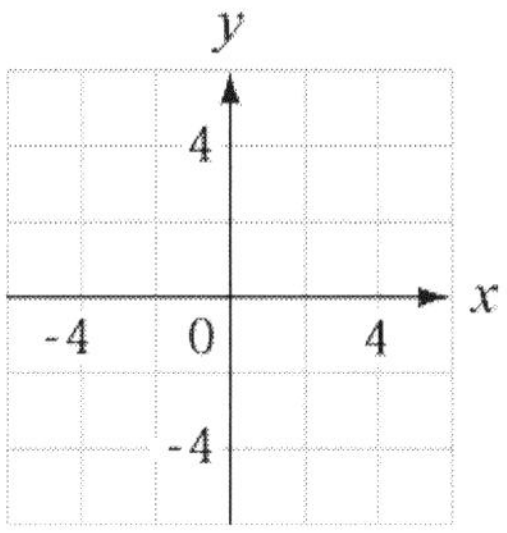


**4.** $y = 2x - 4$

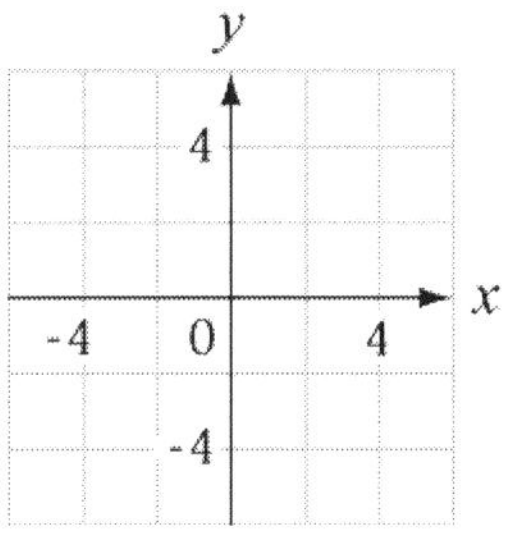


**5.** $3x - y = 3$

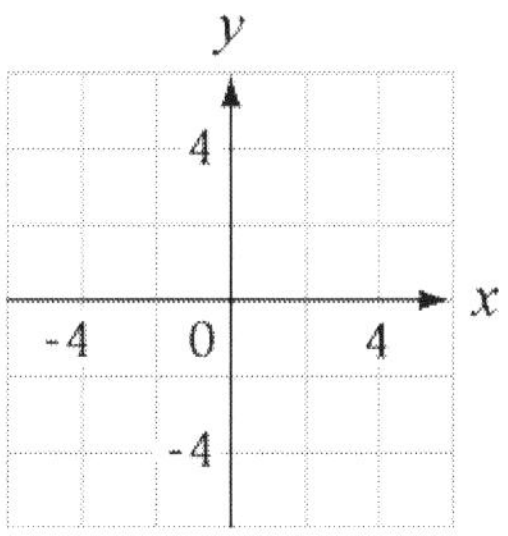


**6.** $2x + y = 4$

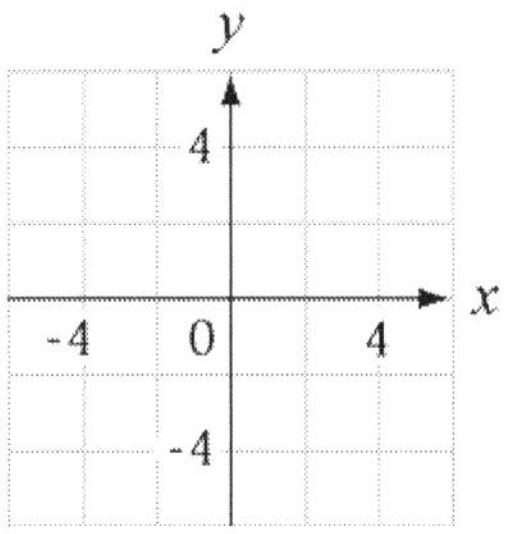


**7.** Graph the line that passes through point (1, 2) and has slope $\frac{1}{3}$.

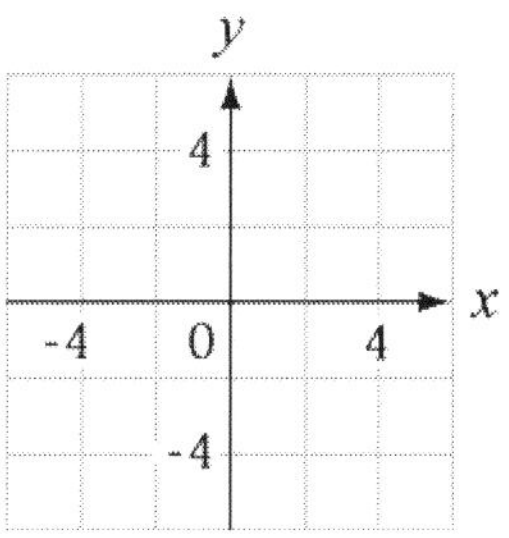


**8.** Graph the line that passes through point (–2, 1) and has slope –1.

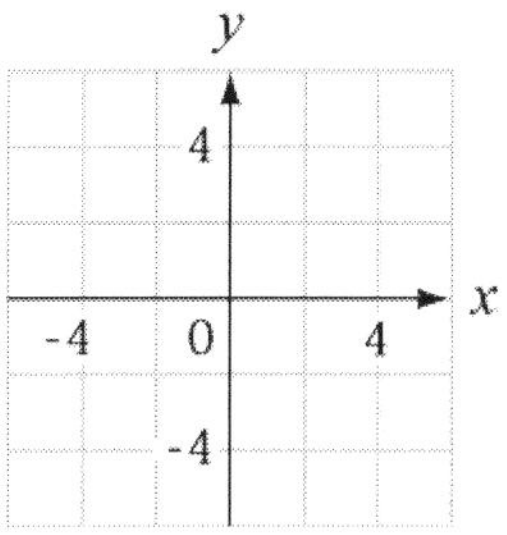

Name Score

**Solve.**

**1.** On April 9 in Sacramento, California, the temperature at 6 A.M. was 44°F. At 1 P.M. on the same day, the temperature was 72°F. Find the average rate of change in temperature per hour.

**2.** In 1971, the cost of a first class stamp was 8 cents. In 1998, the cost of a first class stamp was 32 cents. Find the average rate of change per year.

**1.** ______________

**2.** ______________

The table below shows the number of acres burned by wildfires through September 9 each year. (Source: USA Today)

| Year | 2002 | 2003 | 2004 | 2005 | 2006 |
|---|---|---|---|---|---|
| Acres burned | 6,420,000 | 3,060,000 | 7,610,000 | 7,930,000 | 8,650,000 |

**3.** What was the average annual rate of change in acres burned from 2003 to 2005?

**4.** Was the average annual rate of change in acres burned from 2002 to 2003 greater than or less than the average annual rate of change from 2005 to 2006?

**3.** ______________

**4.** ______________

The table below shows the number of registered lobbyists in Washington D.C. (Source: USA Today)

| Year | 2000 | 2001 | 2002 | 2003 | 2004 | 2005 |
|---|---|---|---|---|---|---|
| Number registered | 16,342 | 18,854 | 21,089 | 24,872 | 30,402 | 32,890 |

**5.** What was the average annual rate of change in the number of registered lobbyists in Washington D.C. from 2001 to 2005?

**6.** Was the average annual rate of change in the number of registered lobbyists in Washington D.C. from 2001 to 2003 greater than or less than the average annual rate of change from 2003 to 2005?

**5.** ______________

**6.** ______________

Name Score

**Find the equation of the line that contains the given point and has the given slope.**

**1.** Point (0, 5). $m = -3$

**2.** Point (–2, 0). $m = 4$

**3.** Point (1, 2). $m = \frac{1}{3}$

**4.** Point (3, –2). $m = -\frac{1}{3}$

**5.** Point (–3, 2). $m = -1$

**6.** Point (–3, 2). $m = \frac{1}{2}$

**7.** Point (–1, 1). $m = \frac{2}{3}$

**8.** Point (0, 3). $m = -\frac{1}{3}$

**9.** Point (–1, 5). $m = -2$

**10.** Point (–1, 3). $m = -\frac{2}{5}$

**11.** Point (1, –2). $m = \frac{3}{4}$

**12.** Point (–1, –3). $m = \frac{1}{3}$

**13.** Point (–1, –1). $m = -\frac{1}{2}$

**14.** Point (0, 0). $m = -\frac{1}{4}$

**15.** Point (3, –2). $m = 2$

**16.** Point (3, –4). $m = -2$

**17.** Point (2, 4). $m = -\frac{1}{3}$

**18.** Point (4, 1). $m = -\frac{3}{5}$

**1.** ____________

**2.** ____________

**3.** ____________

**4.** ____________

**5.** ____________

**6.** ____________

**7.** ____________

**8.** ____________

**9.** ____________

**10.** ____________

**11.** ____________

**12.** ____________

**13.** ____________

**14.** ____________

**15.** ____________

**16.** ____________

**17.** ____________

**18.** ____________

Name Score

**Find the equation of the line through the given points.**

1. $P_1(1, 4)$, $P_2(2, 3)$
2. $P_1(1, 3)$, $P_2(0, 4)$
3. $P_1(1, 2)$, $P_2(4, 4)$
4. $P_1(3, 1)$, $P_2(5, 2)$
5. $P_1(-2, 2)$, $P_2(1, 3)$
6. $P_1(-2, -2)$, $P_2(1, 1)$
7. $P_1(-2, -3)$, $P_2(2, 3)$
8. $P_1(-2, 1)$, $P_2(1, 5)$
9. $P_1(1, 0)$, $P_2(0, -2)$
10. $P_1(0, 3)$, $P_2(-1, 0)$
11. $P_1(0, 0)$, $P_2(2, 1)$
12. $P_1(1, -3)$, $P_2(0, 0)$
13. $P_1(1, 0)$, $P_2(-2, 2)$
14. $P_1(2, -4)$, $P_2(-1, 0)$
15. $P_1(3, -2)$, $P_2(-2, 4)$
16. $P_1(4, -5)$, $P_2(-3, 1)$
17. $P_1(-1, 3)$, $P_2(2, 0)$
18. $P_1(2, -2)$, $P_2(3, -2)$

1. ______________
2. ______________
3. ______________
4. ______________
5. ______________
6. ______________
7. ______________
8. ______________
9. ______________
10. ______________
11. ______________
12. ______________
13. ______________
14. ______________
15. ______________
16. ______________
17. ______________
18. ______________

Name Score

**Solve.**

**1.** A building contractor estimates that the cost to build a home is \$35,000 plus \$95 for each square foot of floor space in the house.
**a.** Determine a linear function that will give the cost of building a house that contains a give number of square feet.
**b.** Use this model to determine the cost to build a house containing 1500 sq. ft.

**2.** The gas tank of a certain car contains 13 gal when the driver of the car begins a trip. Each mile driven by the driver decreases the amount of gas in the tank by 0.025 gal.
**a.** Write a linear function for the number of gallons of gas in the tank in terms of the number of miles driven.
**b.** Use your equation to find the number of gallons in the tank after 180 miles are driven.

**3.** A manufacturer of minivans determined that 75,000 cars per month can be sold at a price of \$24,000. At a price of \$23,500 the number of minivans sold per month would increase to 80,000.
**a.** Determine a linear function that will predict the number of cars that would be sold each month at a given price.
**b.** Use this model to predict the number of cars that would be sold at a price of \$23,250.

**4.** An account executive receives a base salary plus a commission. On \$25,000 in monthly sales, the account executive receives \$2450. On \$40,000 in monthly sales, the account executive receives \$3200.
**a.** Determine a linear function that will yield the compensation of the account executive for a given amount of monthly sales.
**b.** Use this model to determine the account executive's compensation for \$70,000 in monthly sales.

**5.** There are approximately 90 Calories in a 4 oz serving of cottage cheese and approximately 135 Calories in a 6 oz serving.
**a.** Determine a linear function for the number of Calories in a serving of cottage cheese in terms of the size of the serving.
**b.** Use your equation to estimate the number of Calories in a 5 oz serving.

**6.** A cellular phone company offers a plan for people who plan to use the phone only in emergencies. The plan costs the user \$7.95 per month plus \$0.69 per minute used.
**a.** Write a linear function for the monthly cost in terms of the number of minutes used.
**b.** Use your equation to find the cost of using the cellular phone for 9 minutes in one month.

**7.** At sea level, the boiling point of water is 100°C. At an altitude of 1.5 km, the boiling point of water is 94.75°C.
**a.** Write a linear function for the boiling point of water in terms of the altitude above sea level.
**b.** Use your equation to predict the boiling point of water at an altitude of 5 km above sea level.

**8.** Let $f$ be a linear function. If $f(-3)=-1$ and $f(2)=-11$, find $f(x)$.

**1. a.** ______________
**b.** ______________
**2. a.** ______________
**b.** ______________
**3. a.** ______________
**b.** ______________
**4. a.** ______________
**b.** ______________
**5. a.** ______________
**b.** ______________
**6. a.** ______________
**b.** ______________
**7. a.** ______________
**b.** ______________
**8.** ______________

Name Score

**Solve.**

1. Is the line $x = -1$ perpendicular to the line $y = 2$?

2. Is the line $y = \frac{1}{4}$ perpendicular to the line $y = -2$?

3. Is the line $y = 2x - 5$ parallel to the line $y = -\frac{1}{2}x - 5$?

4. Is the line $y = -\frac{1}{3}x + 2$ parallel to the line $y = -\frac{1}{3}x - 1$?

5. Is the line $y = \frac{3}{2}x - 1$ perpendicular to the line $y = -\frac{2}{3}x + 3$?

6. Are the lines $x + 2y = 1$ and $x + 2y = -3$ parallel?

7. Are the lines $x - 3y = 4$ and $3x + y = 6$ perpendicular?

8. Are the line $3x - 2y = 5$ and $3x + 2y = -3$ perpendicular?

9. Is the line that contains the points (4, –5) and (11, –10) perpendicular to the line that contains the points (3, 7) and (–2, 0)?

10. Find the equation of the line containing the point (–2, –5) and parallel to the line $x + 3y = 5$.

11. Find the equation of the line containing the point (3, –1) and perpendicular to the line $y = -4x + 7$.

12. Find the equation of the line containing the point (4, 3) and parallel to the line $5x - 3y = 8$.

13. Find the equation of the line containing the point (3, –2) and perpendicular to the line $y = 2x$.

14. Find the equation of the line containing the point (–1, 4) and perpendicular to the line $y = \frac{3}{4}x - 4$.

15. Find the equation of the line containing the point (–2, –2) and parallel to the line $2x - 3y = 3$.

16. Find the equation of the line containing the point (–1, 5) and parallel to the line $3x + y = -2$.

1. ______
2. ______
3. ______
4. ______
5. ______
6. ______
7. ______
8. ______
9. ______
10. ______
11. ______
12. ______
13. ______
14. ______
15. ______
16. ______

Name Score

**Graph the solution set.**

**1.** $y \geq -x + 2$

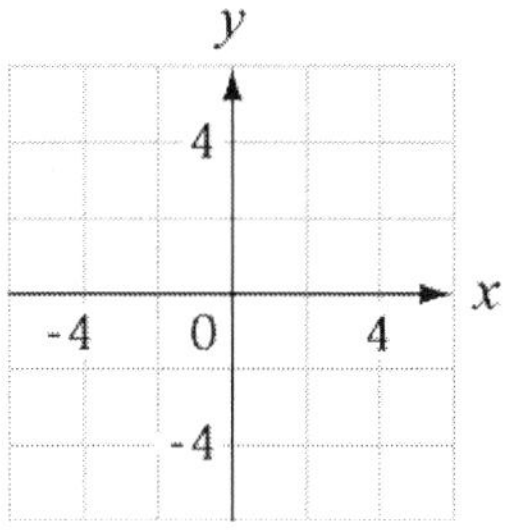


**2.** $y < -\frac{1}{2}x + 2$

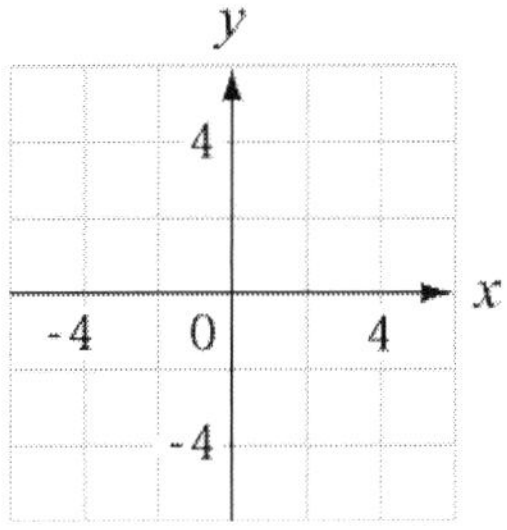


**3.** $x + 3y \leq 6$

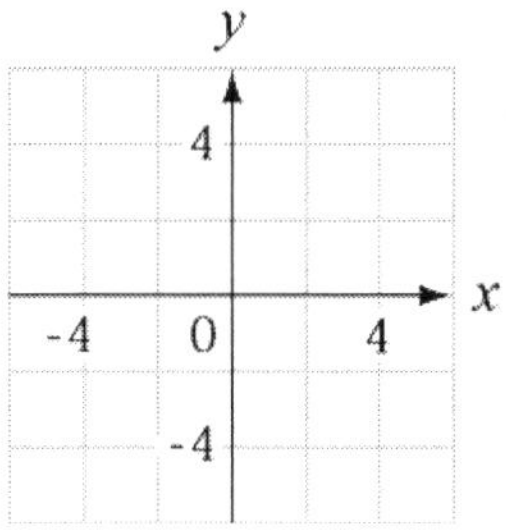


**4.** $3x + 2y \leq 6$

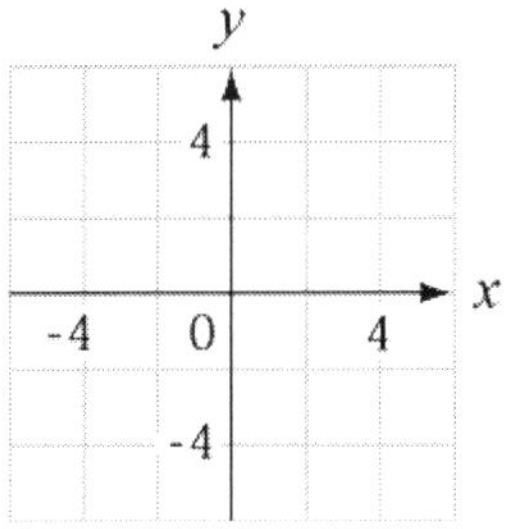


**5.** $5x - 2y < 10$

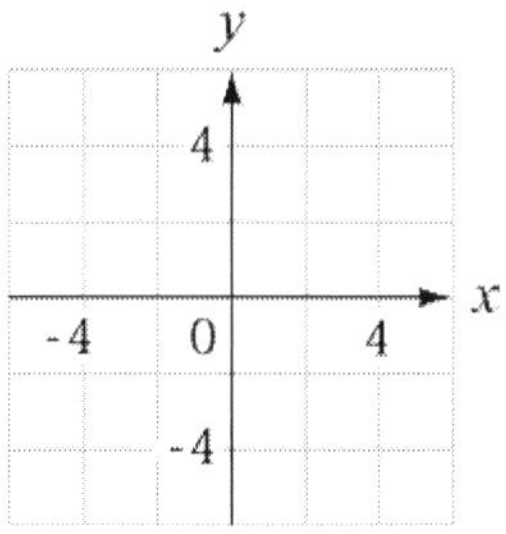


**6.** $3x - 4y \leq 12$

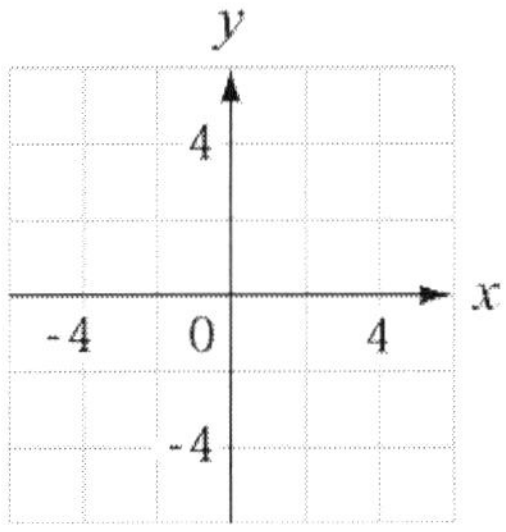


**7.** $y - 3 > 0$

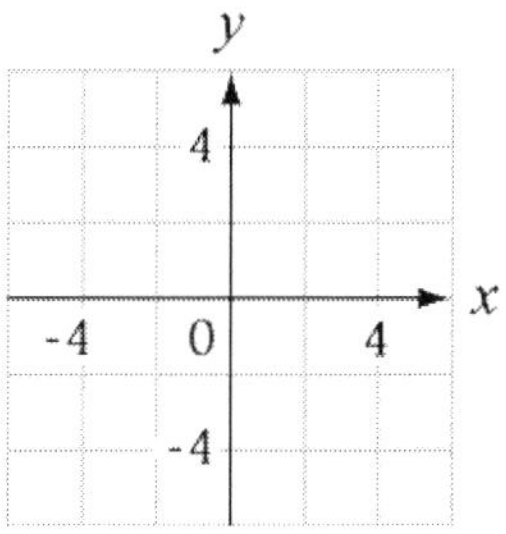


**8.** $x - 1 \geq 0$

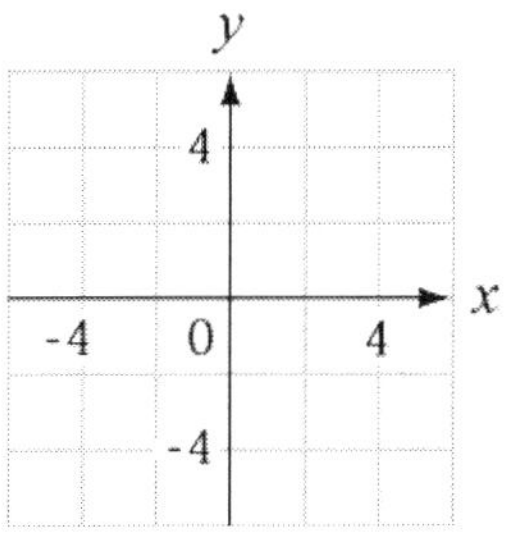


**1.** ____________

**2.** ____________

**3.** ____________

**4.** ____________

**5.** ____________

**6.** ____________

**7.** ____________

**8.** ____________

Name Score

**Solve by graphing.**

1. $x + y = -1$
$x + 2y = 1$

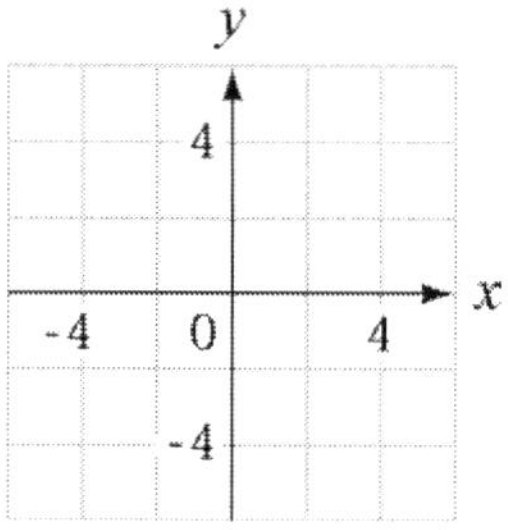


2. $x + y = 1$
$2x - y = 5$

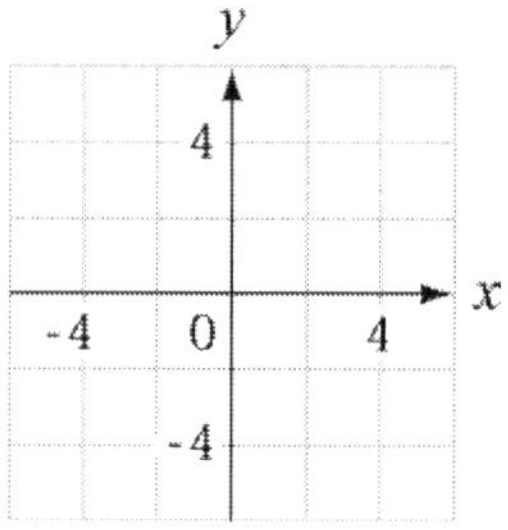


3. $x - y = 1$
$2x - y = -1$

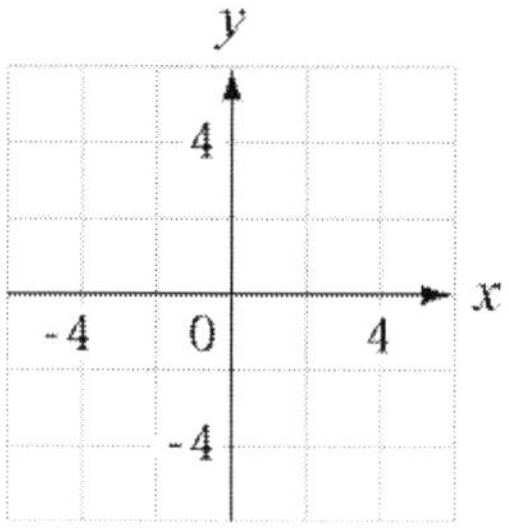


4. $x - 2y = 1$
$x + y = 4$

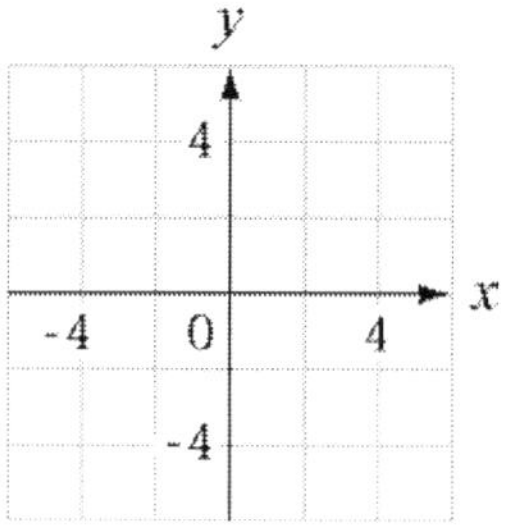


5. $2x - 3y = 10$
$y = -2$

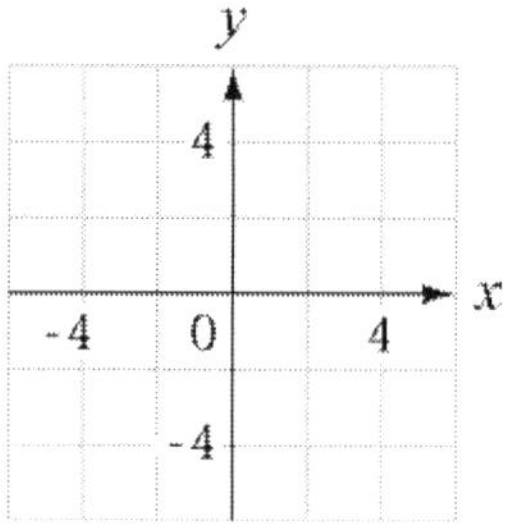


6. $x = 3$
$2x - y = 4$

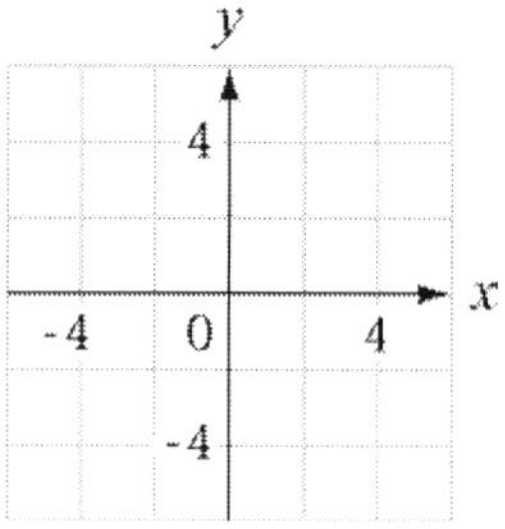


7. $x = 2$
$y = 1$

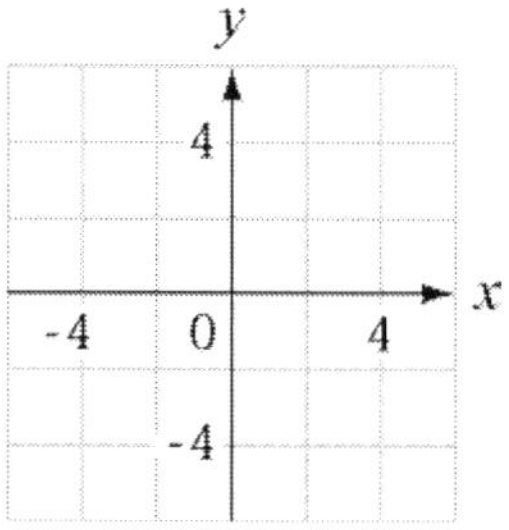


8. $x + 1 = 0$
$y - 3 = 0$

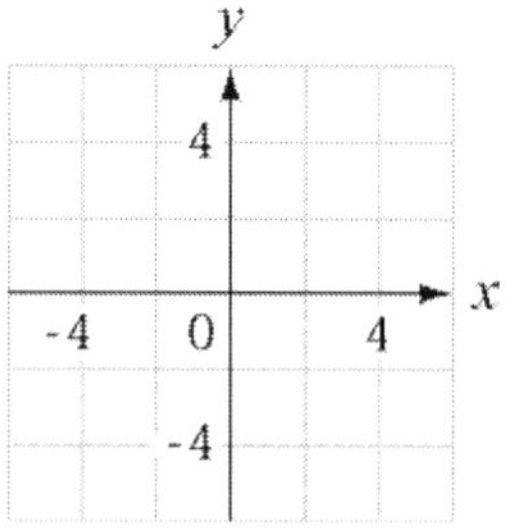

Name Score

**Solve by substitution.**

1. $2x - y = 4$
   $x = 3$

2. $y = -1$
   $3x + 2y = 10$

3. $y = x - 2$
   $2x - y = 5$

4. $y = -x + 2$
   $3x - y = -6$

5. $x = y + 2$
   $x - 2y = -1$

6. $x = 3y - 1$
   $3x - y = 5$

7. $3x - 2y = 7$
   $y = 2x - 4$

8. $4x + y = -5$
   $y = 3x + 9$

9. $2x - 5y = -1$
   $y = 7 - 3x$

10. $5x - 3y = -7$
    $x = y - 1$

11. $3x + 2y = 3$
    $y = 1 - 2x$

12. $5x + 2y = 4$
    $y = x - 5$

13. $3x + y = 1$
    $2x + 2y = 6$

14. $x + 3y = -5$
    $2x - y = -3$

15. $x - 2y = 2$
    $2x - 5y = 3$

16. $4x - 2y = 5$
    $-x + 4y = -3$

17. $2x + 5y = -6$
    $-x + 3y = -8$

18. $3x + 2y = 5$
    $2x + y = 3$

19. $x = 3y + 1$
    $x = 2y - 1$

20. $x = 3y + 2$
    $x = -y - 6$

21. $y = 4x - 1$
    $y = 4 - x$

22. $y = 4 - 3x$
    $y = 3 - 2x$

23. $-x + 2y = 8$
    $4x + y = 13$

24. $2x + y = 3$
    $3x - 2y = -20$

1. ________
2. ________
3. ________
4. ________
5. ________
6. ________
7. ________
8. ________
9. ________
10. ________
11. ________
12. ________
13. ________
14. ________
15. ________
16. ________
17. ________
18. ________
19. ________
20. ________
21. ________
22. ________
23. ________
24. ________

Name Score

**Solve by the addition method.**

1. $x - y = 3$
   $x + y = 5$

2. $x + y = 5$
   $3x - y = 7$

3. $x - y = 7$
   $3x + y = 9$

4. $x - 2y = 3$
   $x + 3y = -2$

5. $2x + y = 1$
   $x + 2y = 5$

6. $x - 4y = -10$
   $3x - 2y = -10$

7. $2x - 3y = 7$
   $3x + y = 16$

8. $2x - y = 1$
   $3x - 2y = -1$

9. $x + 3y = -2$
   $2x - y = 10$

10. $x + 2y = 12$
    $3x - 2y = -4$

11. $2x - 3y = 7$
    $4x - 6y = 14$

12. $2x - y = 4$
    $6x - 3y = 12$

13. $8x - y = 3$
    $4x + 2y = 4$

14. $3x + 7y = 1$
    $4x + 5y = -3$

15. $6x + 2y = 0$
    $2x - 5y = 17$

16. $2x + 11y = -1$
    $3x + 2y = 13$

17. $2x + 3y = -8$
    $5x - 6y = 7$

18. $2x - 5y = 19$
    $3x - 2y = 12$

19. $3x - 4y = -11$
    $2x - 3y = -8$

20. $4x - 11y = 38$
    $5x + 2y = 16$

21. $2x - 3y = -7$
    $3x + 2y = -4$

22. $4x + 3y = 2x + 5$
    $3x - 2y = x + 10$

23. $5x + y = 2x - 4$
    $4x - 2y = 3x + 7$

24. $\frac{3}{5}x - \frac{1}{3}y = 7$
    $\frac{1}{4}x - \frac{2}{5}y = -1$

1. ____________
2. ____________
3. ____________
4. ____________
5. ____________
6. ____________
7. ____________
8. ____________
9. ____________
10. ____________
11. ____________
12. ____________
13. ____________
14. ____________
15. ____________
16. ____________
17. ____________
18. ____________
19. ____________
20. ____________
21. ____________
22. ____________
23. ____________
24. ____________

Name Score

**Solve by the addition method.**

**1.**
$$\begin{aligned} 2x+2y+z&=1\\ x-y+z&=0\\ x+y-z&=-4 \end{aligned}$$

**2.**
$$\begin{aligned} x+y-z&=1\\ 2x+3y+z&=3\\ x-y-z&=5 \end{aligned}$$

**3.**
$$\begin{aligned} x+2y-z&=8\\ -x+2y+z&=9\\ x+y+z&=-4 \end{aligned}$$

**4.**
$$\begin{aligned} 2x-y+z&=-9\\ 3x+y-2x&=-7\\ x+4y+3z&=16 \end{aligned}$$

**5.**
$$\begin{aligned} 2x+y+3z&=8\\ x+y+z&=1\\ 4x-y-z&=14 \end{aligned}$$

**6.**
$$\begin{aligned} x+y-z&=-2\\ 2x-y+z&=-1\\ x+2y-3z&=-7 \end{aligned}$$

**7.**
$$\begin{aligned} 3x-y+z&=4\\ 2x-3y+2z&=6\\ x-y+z&=2 \end{aligned}$$

**8.**
$$\begin{aligned} 2x-y-3z&=-4\\ x+y+z&=0\\ x-y+4z&=-9 \end{aligned}$$

**9.**
$$\begin{aligned} 4x-y+3z&=35\\ x-3y-2z&=9\\ 3x+2y+6z&=20 \end{aligned}$$

**10.**
$$\begin{aligned} 3x+4y-2z&=-1\\ 5x+5y-3z&=2\\ 6x+3y-2z&=5 \end{aligned}$$

**1.** ______________

**2.** ______________

**3.** ______________

**4.** ______________

**5.** ______________

**6.** ______________

**7.** ______________

**8.** ______________

**9.** ______________

**10.** ______________

Name Score

**Evaluate the determinant.**

1. $\begin{vmatrix} 4 & 1 \\ -2 & 2 \end{vmatrix}$

2. $\begin{vmatrix} 2 & 5 \\ 3 & 4 \end{vmatrix}$

3. $\begin{vmatrix} 6 & -9 \\ 2 & -1 \end{vmatrix}$

4. $\begin{vmatrix} 2 & 3 \\ 1 & -4 \end{vmatrix}$

5. $\begin{vmatrix} 4 & 3 \\ 5 & -1 \end{vmatrix}$

6. $\begin{vmatrix} 2 & 1 \\ -3 & -8 \end{vmatrix}$

7. $\begin{vmatrix} 5 & 1 \\ -4 & 6 \end{vmatrix}$

8. $\begin{vmatrix} -2 & 3 \\ 1 & -2 \end{vmatrix}$

9. $\begin{vmatrix} 5 & -3 \\ 4 & -2 \end{vmatrix}$

10. $\begin{vmatrix} 3 & 4 & -1 \\ 2 & -1 & 6 \\ 1 & 2 & 3 \end{vmatrix}$

11. $\begin{vmatrix} 5 & 1 & 5 \\ -3 & 1 & 2 \\ 1 & 2 & 4 \end{vmatrix}$

12. $\begin{vmatrix} 2 & 5 & -2 \\ 3 & -2 & 5 \\ -2 & -3 & 2 \end{vmatrix}$

13. $\begin{vmatrix} 1 & 4 & 2 \\ 2 & 3 & 1 \\ 6 & 2 & 4 \end{vmatrix}$

14. $\begin{vmatrix} 2 & 1 & 4 \\ 5 & 6 & 5 \\ -1 & 2 & -1 \end{vmatrix}$

15. $\begin{vmatrix} 1 & 3 & 4 \\ 1 & 3 & 2 \\ -1 & 5 & 2 \end{vmatrix}$

16. $\begin{vmatrix} 2 & -2 & 7 \\ 1 & -1 & 3 \\ 3 & 4 & 2 \end{vmatrix}$

17. $\begin{vmatrix} 1 & 5 & -2 \\ 2 & -3 & 3 \\ 1 & 3 & 3 \end{vmatrix}$

18. $\begin{vmatrix} 1 & 0 & 2 \\ 3 & -1 & 3 \\ 2 & 2 & 8 \end{vmatrix}$

1. ______
2. ______
3. ______
4. ______
5. ______
6. ______
7. ______
8. ______
9. ______
10. ______
11. ______
12. ______
13. ______
14. ______
15. ______
16. ______
17. ______
18. ______

Name Score

**Solve by using Cramer's Rule.**

1. $2x + y = 10$
   $3x - y = -5$

2. $5x - 3y = 3$
   $2x + y = -1$

3. $2x - 3y = 5$
   $5x - 2y = 18$

4. $x - 3y = 6$
   $3x + 5y = 4$

5. $3x - 2y = -5$
   $8x - 3y = 3$

6. $2x + 5y = -3$
   $3x - 7y = 10$

7. $6x - 5y = 2$
   $9x + 10y = 38$

8. $2x + 5y = -31$
   $8x - 3y = -9$

9. $3x + 5y = 9$
   $9x - 2y = -7$

10. $5x + 3y = 6$
    $4x - 2y = -4$

11. $x + 3y - z = -2$
    $2x - y + z = 1$
    $x + 2y - z = -3$

12. $3x + y - z = 0$
    $2x + 2y - z = 1$
    $x + 3y + z = 12$

13. $x + 3y - z = -2$
    $2x + y - 2z = 1$
    $3x - y - 2z = 5$

14. $x + 3y + z = 5$
    $3x - y + 2z = 2$
    $2x + y - z = -4$

15. $5x - 3y + z = -10$
    $-3x + 4y - z = 10$
    $2x + 2y + 3z = 5$

16. $3x - 2y + 4z = 7$
    $5x - 3y - z = 12$
    $5x + 3y + 5z = 15$

1. ______________
2. ______________
3. ______________
4. ______________
5. ______________
6. ______________
7. ______________
8. ______________
9. ______________
10. ______________
11. ______________
12. ______________
13. ______________
14. ______________
15. ______________
16. ______________

Name Score

**Write each matrix in row echelon form.**

**1.** $\left[\begin{array}{cc|c} 4 & -5 & 3 \\ 1 & -3 & 6 \end{array}\right]$

**2.** $\left[\begin{array}{cc|c} 7 & -1 & 2 \\ -6 & 1 & 5 \end{array}\right]$

1. ______

2. ______

**Solve by using the Gaussian elimination method.**

**3.** $\begin{aligned} 2x - 3y &= -6 \\ x + 6y &= 2 \end{aligned}$

**4.** $\begin{aligned} 2x - y &= -4 \\ 4x - 3y &= -5 \end{aligned}$

3. ______

4. ______

**5.** $\begin{aligned} 2x + y &= -2 \\ 3x + 5y &= 11 \end{aligned}$

**6.** $\begin{aligned} 3x - 4y &= -1 \\ 5x + 6y &= 30 \end{aligned}$

5. ______

6. ______

**7.** $\begin{aligned} x + y + 2z &= 2 \\ 3x + 2y + 3z &= -3 \\ 2x + 2y + 3z &= 2 \end{aligned}$

**8.** $\begin{aligned} x + y - z &= -2 \\ 2x + y - 2z &= -5 \\ -3x - y + 3z &= 10 \end{aligned}$

7. ______

8. ______

**9.** $\begin{aligned} x - y &= 2 \\ 2x + y - z &= 3 \\ 2y - z &= 2 \end{aligned}$

**10.** $\begin{aligned} x - y + z &= 6 \\ 3x - y - 2x &= 9 \\ x + 2y + 3z &= 2 \end{aligned}$

9. ______

10. ______

Name Score

**Solve.**

1. Flying with the wind, a small plane flew 615 mi in 3 h. Against the wind, the plane could fly only 405 mi in the same amount of time. Find the rate of the plane in calm air and the rate of the wind.

2. A jet plane flying with the wind went 2750 mi in 5 h. Against the wind, the plane could fly only 2450 mi in the same amount of time. Find the rate of the plane in calm air and the rate of the wind.

3. A cabin cruiser traveling with the current went 60 mi in 3 h. Against the current, it took 6 h to travel the same distance. Find the rate of the cabin cruiser in calm water and the rate of the current.

4. A motorboat traveling with the current went 72 mi in 4 h. Against the current, it took 6 h to travel the same distance. Find the rate of the boat in calm water and the rate of the current.

5. Flying with the wind, a pilot flew 800 mi between two cities in 4 h. The return trip against the wind took 5 h. Find the rate of the plane in calm air and the rate of the wind.

6. A corporate jet flying with the wind flew 1000 mi in 4 h. Flying against the wind, the plane required 5 h to travel the same distance. Find the rate of the plane in calm air.

7. A motorboat traveling with the current went 112 km in 4 h. Against the current, the boat could go only 80 km in the same amount of time. Find the rate of the boat in calm water and the rate of the current.

8. A rowing team rowing with the current traveled 45 mi in 3 h. Against the current, the team rowed 27 mi in 3 h. Find the rate of the rowing team in calm water and the rate of the current.

9. A plane flying with a tailwind flew 480 mi in 3 h. Against the wind, the plane required 4 h to fly the same distance. Find rate of the plane in calm air and the rate of the wind.

10. Flying with the wind, a plane flew 1080 mi in 3 h. Against the wind, the plane required 4 h to fly the same distance. Find the rate of the plane in calm air and the rate of the wind.

1. ______________
2. ______________
3. ______________
4. ______________
5. ______________
6. ______________
7. ______________
8. ______________
9. ______________
10. ______________

Name Score

**Solve.**

**1.** A carpenter purchased 50 ft of redwood and 90 ft of pine for a total cost of $33. A second purchase, at the same prices, included 80 ft of redwood and 70 ft of pine for a total cost of $38. Find the cost per foot of redwood and of pine.

**2.** A merchant mixed 15 lb of a cinnamon tea with 10 lb of spice tea. The 25-pound mixture sells for $65. A second mixture included 14 lb of the cinnamon tea and 16 lb of the spice tea. The 30-pound mixture sells for $84. Find the cost per pound for the cinnamon and for the spice tea.

**3.** During one month, a small business used 900 units of electricity and 200 units of gas for a total cost of $218. The next month 800 units of electricity and 250 units of gas were used for a total cost of $201. Find the cost per unit of gas.

**4.** A contractor buys 20 yd of nylon carpet and 30 yd of wool carpet for $860. A second purchase, at the same prices, includes 24 yd of nylon carpet and 32 yd of wool carpet for $984. Find the cost per yard of the wool carpet.

**5.** The total value of the quarters and nickels in a coin bank is $14.00. If the quarters were nickels and the nickels were quarters, the total value of the coins would be $22.00. Find the number of quarters in the bank.

**6.** A coin bank contains only quarters and dimes. The total value of the coins in the bank is $8.80. If the dimes were quarters and the quarters were dimes, the total value of the coins would be $7.30. Find the number of dimes in the bank.

**7.** A restaurant manager buys 120 lb of hamburger and 60 lb of steak for a total cost of $360. A second purchase, at the same prices, includes 200 lb of hamburger and 80 lb of steak. The total cost is $540. Find the cost of 1 lb of steak.

**8.** A sheet metal shop ordered 60 lb of tin and 30 lb of a zinc alloy for a total cost of $480. A second purchase, at the same prices, included 40 lb of tin and 70 lb of the zinc alloy. The total cost was $770. Find the total cost per pound of the tin and of the zinc alloy.

**9.** A citrus fruit grower purchased 30 orange trees and 35 grapefruit trees for $450. The next week, at the same prices, the grower bought 40 orange trees and 40 grapefruit trees for $560. Find the cost of an orange tree and of a grapefruit tree.

**10.** A store owner purchased thirty 75-watt light bulbs and 40 fluorescent light bulbs for a total cost of $95. A second purchase, at the same prices, included forty 75-watt light bulbs and 20 fluorescent lights for a total cost of $60. Find the cost of a 75-watt bulb and of a fluorescent light.

**1.** ______
**2.** ______
**3.** ______
**4.** ______
**5.** ______
**6.** ______
**7.** ______
**8.** ______
**9.** ______
**10.** ______

Name Score

**Graph the solution set.**

1. $y \le x + 2$
$y \ge -x + 3$

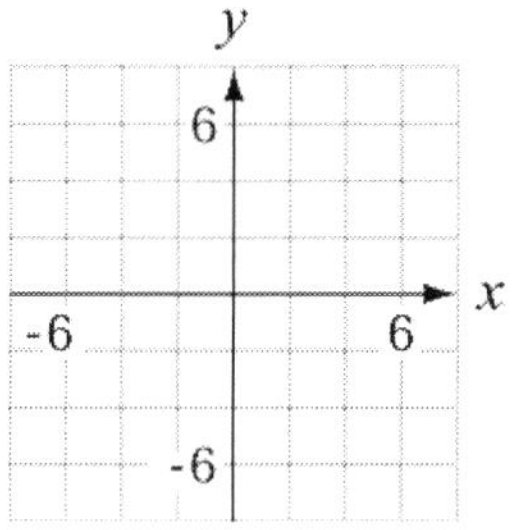


2. $y \ge 2x + 3$
$y \ge -x + 2$

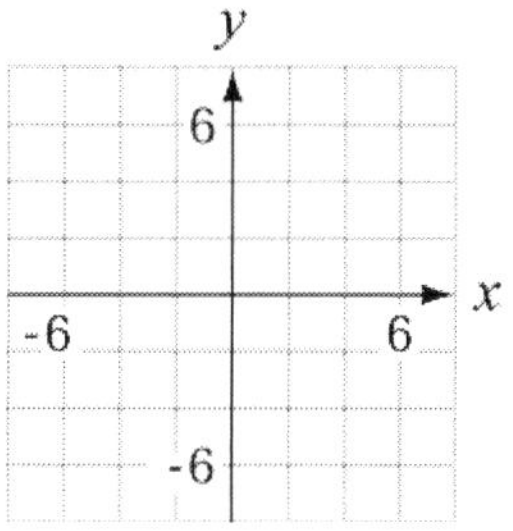


3. $x - 3y \le 6$
$2x + y \le 4$

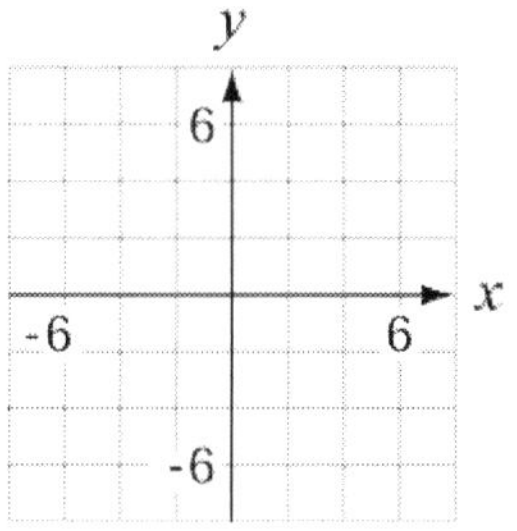


4. $x + 2y \le 4$
$x - y \le 3$
$y \ge 0$

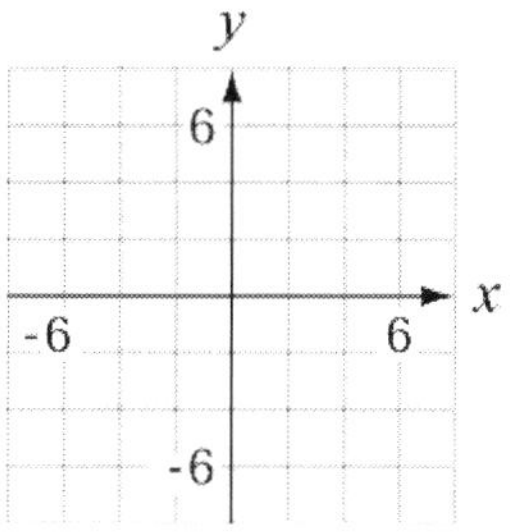


5. $x - 2y \ge 1$
$2x + y \ge 2$

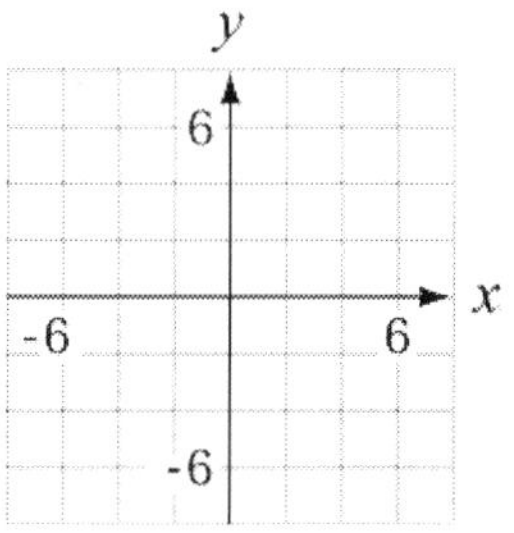


6. $x - 3y \le 0$
$2x - y \ge -3$

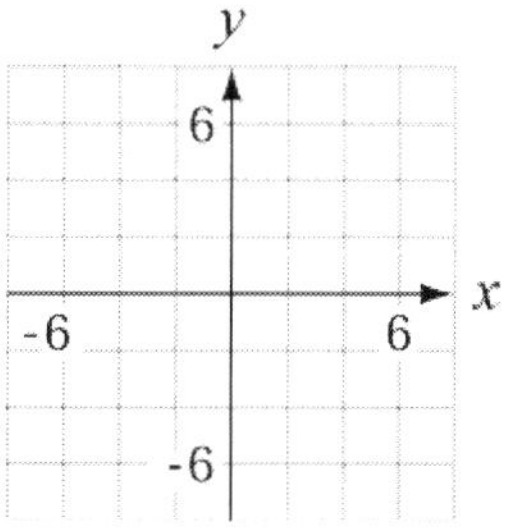


7. $2x - y \le 0$
$x - 2y \ge -2$

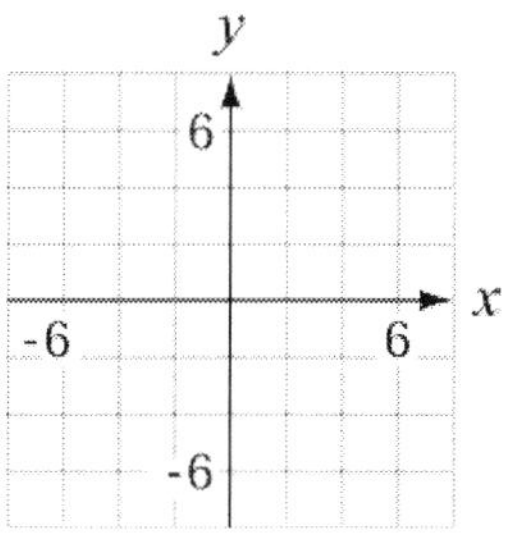


8. $x - 3y \ge -3$
$x - 3y \le 6$

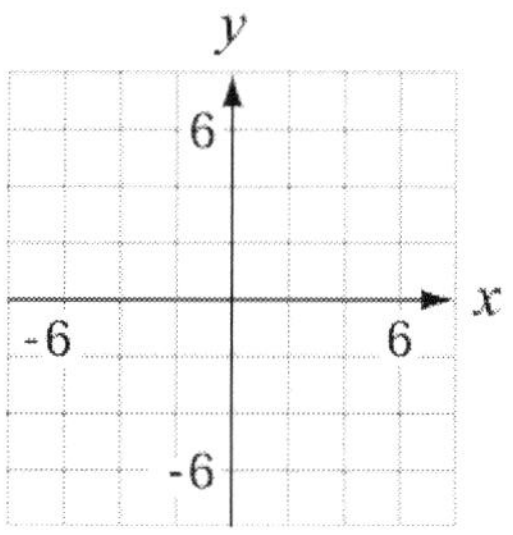

Name Score

**Evaluate.**

1. Given $P(x) = 4x^3 - 3x^2 + 6$, evaluate $P(-3)$.

2. Given $P(x) = 4x^3 - 3x^2 + 6$, evaluate $P(-3)$.

1. ______________

2. ______________

**Indicate which define a polynomial function For those that are polynomial functions: a. Identify the leading coefficient. b. Identify the constant term. c. State the degree.**

3. $P(x) = 6x^4 - 2x + 7$

4. $R(x) = \dfrac{x^2 + 2x + 1}{x}$

3. a. ______________
b. ______________
c. ______________

4. a. ______________
b. ______________
c. ______________

5. $f(x) = x^2 - 6x - \sqrt{7}$

6. $g(x) = \pi$

5. a. ______________
b. ______________
c. ______________

6. a. ______________
b. ______________
c. ______________

**Graph.**

7. $f(x) = x^2 - 3x + 2$

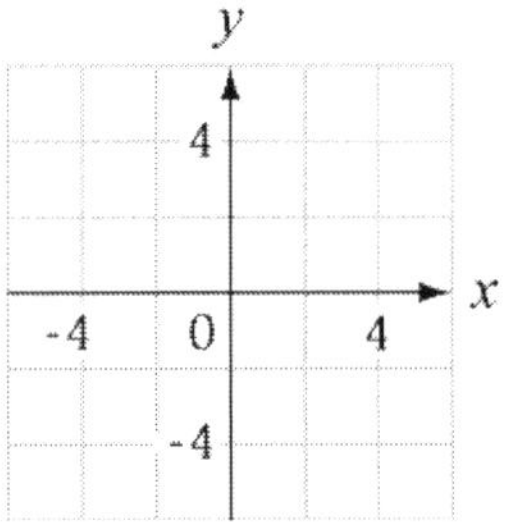


8. $f(x) = 2x^2 + 1$

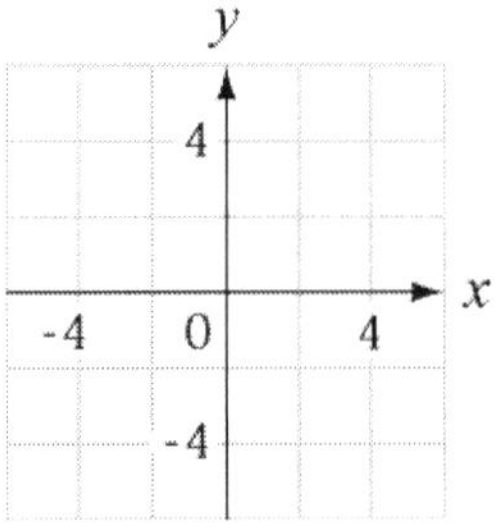


9. $f(x) = 2x^2 - x - 3$

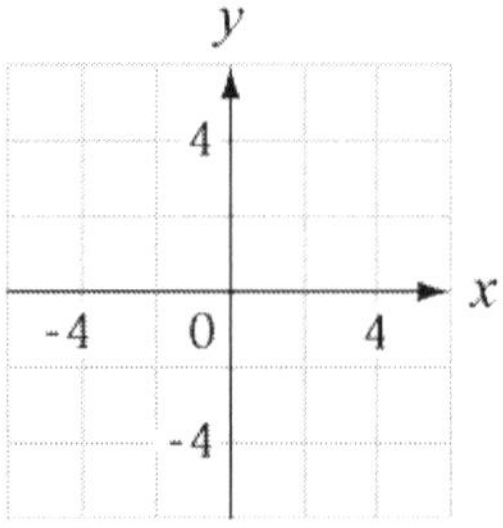

Name Score

**Simplify.**

**1.** Given $P(x) = x^2 - 6$ and $R(x) = 4x - 2$, find $P(x) - R(x)$.

**2.** Given $P(x) = 3x^3 - 4x^2 - 6x$ and $R(x) = -6x^2 + 10x + 4$, find $P(x) - R(x)$.

**1.** ______________

**2.** ______________

**Simplify. Use a vertical format.**

**3.** $(x^2 - 4xy + 2y^2) + (3x^2 - 4y^2)$

**4.** $(2x^2 + 3y^2) + (-4x^2 + 3xy - 2y^2)$

**3.** ______________

**4.** ______________

**5.** $(2x^{2n} + 6x^n - 4) + (-x^{2n} - x^n + 7)$

**6.** $(x^{2n} - 2x^n - 2) + (4x^{2n} + 6n + 3)$

**5.** ______________

**6.** ______________

**Simplify. Use a horizontal format.**

**7.** $(4y^3 - 8y) + (y^2 - 7y + 1)$

**8.** $(-3y^2 - 5y - 11) + (6y^2 - 3y)$

**7.** ______________

**8.** ______________

**9.** $(3a^2 - 2a - 6) - (-3a^2 - a - 8)$

**10.** $(4x^4 - 2x^3 + x^2) + (4x^3 - 6x^2 + 3x)$

**9.** ______________

**10.** ______________

**11.** $(2x^4 - 3x + 2) + (4x^3 - 4x - 7)$

**12.** $(a^{2n} - 3a^n - 4) - (3a^{2n} - 2a^n + 7)$

**11.** ______________

**12.** ______________

**13.** $(x^{2n} - 2x^n - 5) - (3x^{2n} - x^n + 6)$

**14.** $(3x^2 - 2x - 8) - (-x^3 - 4x^2 + 5x)$
$-(4x^3 - 2x^2 - 8x + 3)$

**13.** ______________

**14.** ______________

**15.** $(3a^3 - 3a^2b + 4ab^2 + 5b^3) - (5a^2b - 4ab^2 - 6b^3)$

**15.** ______________

Name Score

**Simplify.**

1. $(x)(3x)$
2. $(-2y)(y)$
3. $(2x)(3x)$
4. $(4y^3)(4y^2)$
5. $(-2a^2)(-4a^3)$
6. $(4a^6)(-2a^4)$
7. $(xy^2)(x^2y^3)$
8. $(-3x^3)(4x^2y)$
9. $(-3a^3)(3a^3b^2)$
10. $(ab^3)(a^2b^2)$
11. $(4xy^2)(-2x^2y^3)$
12. $(-5ab^2)(-a^2b^3)$
13. $(xy^2z)(x^2y^3)$
14. $(ab^2)(a^2bc^3)$
15. $(xy^2z)(x^3y^2)$
16. $(-ab^2)(a^3b^5)$
17. $(x^2y^3)(-x^2y)$
18. $(-4y^3z)(-7y^4z^3)$
19. $(2x^2y)(-3x^2y)$
20. $(-a^2bc^2)(ab^2c)(-abc^3)$
21. $(x^2y^2z)(x^2y^3)(y^3z^2)$
22. $(-2ab)(3a^2b^2)(-4a^3b^3)$
23. $(2x^2y^4)(4xy)(-3x^2y^2)$
24. $(2a^2b^2)(-3ab^2)(a^4b^2)$
25. $(x^2y^3)(-2xy)(-4x^2y^4)$
26. $(3a^2b)(-2a^3b^2)(a^5b)$
27. $(2ab^2)(-3abc^2)(5a^2c)$

1. __________
2. __________
3. __________
4. __________
5. __________
6. __________
7. __________
8. __________
9. __________
10. __________
11. __________
12. __________
13. __________
14. __________
15. __________
16. __________
17. __________
18. __________
19. __________
20. __________
21. __________
22. __________
23. __________
24. __________
25. __________
26. __________
27. __________

Name Score

**Simplify.**

1. $(x^2)^2$
2. $(x^2)^3$
3. $(y^2)^4$
4. $(x^5)^2$
5. $(y^4)^3$
6. $(-x^3)^2$
7. $(2x)^3$
8. $(3y)^2$
9. $(-2x)^3$
10. $(-3y^2)^2$
11. $(xy^2)^2$
12. $(x^2y^3)^4$
13. $(-2ab^2)^3$
14. $(x^2y^2)^5$
15. $(xy^3)^5$
16. $(x^4y^3)^4$
17. $(2x^2y^3)^2$
18. $(-a^2b^4)^3$
19. $(a^2b)^2(ab^2)^2$
20. $(-2y)(3y^2)^3$
21. $(a^2b)(ab)^2$
22. $(-3x)(-2x^2y^2)^2$
23. $(-2y)(-3x^2y)^2$
24. $(ab^2)(-2ab^3)^2$
25. $(-2a^2)(3ab^2)^3$
26. $(2a^2b)^3(-3b^2)$
27. $(-3ab^2)^3(-3ab)^2$

1. ______
2. ______
3. ______
4. ______
5. ______
6. ______
7. ______
8. ______
9. ______
10. ______
11. ______
12. ______
13. ______
14. ______
15. ______
16. ______
17. ______
18. ______
19. ______
20. ______
21. ______
22. ______
23. ______
24. ______
25. ______
26. ______
27. ______

Name Score

**Multiply.**

1. $x(x+1)$
2. $y(2-y)$
3. $-x(x+2)$
4. $2a(a-1)$
5. $3b(b+5)$
6. $-2x^2(x-1)$
7. $-4y^2(y+6)$
8. $-x^2(2x^2-3)$
9. $2x(5x^2-2x)$
10. $3y(2y-y^2)$
11. $(2x-3)x$
12. $(2x-1)3x$
13. $-x^2y(x-y^2)$
14. $-xy^2(2x-y)$
15. $x(x^2-2x+1)$
16. $x(x^2+3x-2)$
17. $-a(a^2-6a-1)$
18. $-b(2b^2+3b-6)$
19. $x^2(2x^2-3x-2)$
20. $y^3(-2y^2-3y+4)$
21. $2y^2(-2y^2-5y+8)$
22. $3x^2(4x^2-2x+7)$
23. $4x^2(5x^2-x-9)$
24. $5y^2(-y^2+3y-6)$
25. $ab(a^2-3ab-4b^2)$
26. $xy(x^2-2xy+2y^2)$
27. $ab(a^2+5ab-7b^2)$

1. ________
2. ________
3. ________
4. ________
5. ________
6. ________
7. ________
8. ________
9. ________
10. ________
11. ________
12. ________
13. ________
14. ________
15. ________
16. ________
17. ________
18. ________
19. ________
20. ________
21. ________
22. ________
23. ________
24. ________
25. ________
26. ________
27. ________

Name Score

**Multiply.**

1. $(x^2+2x+1)(x+2)$
2. $(x^2-3x+5)(x-1)$
3. $(a^2-a+2)(a-2)$
4. $(x^2-4x-5)(x-3)$
5. $(a^2-5a+3)(a-1)$
6. $(x^2-5x-4)(3x-1)$
7. $(-2b^2-b+2)(b-4)$
8. $(-a^2+4a-3)(2a+3)$
9. $(-x^2+6x-3)(2x-3)$
10. $(-a^2-a+2)(3a+1)$
11. $(x^2-4)(x+1)$
12. $(y^2+3y)(y-3)$
13. $(x^2+3)(x-5)$
14. $(y^2-y)(2y+4)$
15. $(x^3-2x+1)(x-3)$
16. $(y^3+2y^2-5)(2y-3)$
17. $(4y^2+7y-1)(2y-7)$
18. $(3y^2+y-3)(3y-2)$
19. $(3a^3-2a+3)(a-2)$
20. $(2y^2+4y-2)(2y-1)$
21. $(4a^3-4a+1)(a-3)$
22. $(2b^3-3b^2+6)(3b-1)$
23. $(y^3+y^2-2y+2)(y+1)$
24. $(a^3-2a^2+3a-1)(3a-2)$

1. ____________
2. ____________
3. ____________
4. ____________
5. ____________
6. ____________
7. ____________
8. ____________
9. ____________
10. ____________
11. ____________
12. ____________
13. ____________
14. ____________
15. ____________
16. ____________
17. ____________
18. ____________
19. ____________
20. ____________
21. ____________
22. ____________
23. ____________
24. ____________

Name Score

**Multiply.**

1. $(y+2)(y+3)$
2. $(a-2)(a+5)$
3. $(b-5)(b+4)$
4. $(y+2)(y-7)$
5. $(x+9)(x-4)$
6. $(y-6)(y-2)$
7. $(a-2)(a-8)$
8. $(x+11)(x-3)$
9. $(2x+1)(x+6)$
10. $(y+1)(3y+2)$
11. $(2x-3)(x+3)$
12. $(5x-2)(x+3)$
13. $(2x-1)(3x-5)$
14. $(2y-9)(y+1)$
15. $(4y-7)(y+2)$
16. $(5a+2)(6a+1)$
17. $(6a-13)(2a-5)$
18. $(5a-9)(2a-7)$
19. $(3b+11)(5b-4)$
20. $(3a+10)(4a-3)$
21. $(x+y)(x+2y)$
22. $(2a+b)(a+2b)$
23. $(2x-3y)(x-y)$
24. $(a-3b)(2a+3b)$
25. $(4a-b)(2a+5b)$
26. $(3x-5y)(3x+2y)$
27. $(5x+2y)(6x+y)$

1. __________
2. __________
3. __________
4. __________
5. __________
6. __________
7. __________
8. __________
9. __________
10. __________
11. __________
12. __________
13. __________
14. __________
15. __________
16. __________
17. __________
18. __________
19. __________
20. __________
21. __________
22. __________
23. __________
24. __________
25. __________
26. __________
27. __________

Name Score

**Simplify.**

1. $(x+1)(x-1)$
2. $(x-3)(x+3)$
3. $(x-7)(x+7)$
4. $(2x-1)(2x+1)$
5. $(3x-1)(3x+1)$
6. $(x+5)^2$
7. $(y-4)^2$
8. $(3y-1)^2$
9. $(x-1)^2$
10. $(x+7)^2$
11. $(x+9)^2$
12. $(x-y)^2$
13. $(2a-5)^2$
14. $(5x-4)^2$
15. $(3x-7)^2$
16. $(3a-5)(3a+5)$
17. $(2x+5)^2$
18. $(9x-2)^2$
19. $(a-2b)^2$
20. $(x+2y)^2$
21. $(5x-6)(5x+6)$
22. $(b-6a)(b+6a)$
23. $(x+5y)^2$
24. $(2-7y)^2$
25. $(3-5y)(3+5y)$
26. $(4x-1)(4x+1)$
27. $(x+6y)^2$

1. ______
2. ______
3. ______
4. ______
5. ______
6. ______
7. ______
8. ______
9. ______
10. ______
11. ______
12. ______
13. ______
14. ______
15. ______
16. ______
17. ______
18. ______
19. ______
20. ______
21. ______
22. ______
23. ______
24. ______
25. ______
26. ______
27. ______

Name Score

**Solve.**

1. The length of a rectangle is $3x$ m. The width is $(3x - 1)$ m. Find the area of the rectangle in terms of the variable $x$.

2. The width of a rectangle is $(x - 2)$ in. The length is $(3x + 2)$ in. Find the area of the rectangle in terms of the variable $x$.

3. The length of a rectangle is $(3x + 1)$ ft. The width is $(2x - 1)$ ft. Find the area of the rectangle in terms of the variable $x$.

4. The length of a side of a square is $(x + 3)$ cm. Use the equation $A = s^2$, where $s$ is the length of a side of a square, to find the area of the square in terms of the variable $x$.

5. The length of a side of a square is $(x - 8)$ km. Use the equation $A = s^2$, where $s$ is the length of a side of a square, to find the area of the square in terms of the variable $x$.

6. The length of a side of a square is $(2x + 1)$ in. Find the area of the square in terms of the variable $x$.

7. The length of a side of a square is $(3x - 4)$ ft. Find the area of the square in terms of the variable $x$.

8. The radius of a circle is $(x - 3)$ m. Use the equation $A = \pi r^2$, where $r$ is the radius, to find the area of the circle in terms of the variable $x$. Leave the answer in terms of $\pi$.

9. The radius of a circle is $(x + 6)$ in. Use the equation $A = \pi r^2$, where $r$ is the radius, to find the area of the circle in terms of the variable $x$. Leave the answer in terms of $\pi$.

10. The radius of a circle is $(2x + 1)$ ft. Use the equation $A = \pi r^2$, where $r$ is the radius, to find the area of the circle in terms of the variable $x$. Leave the answer in terms of $\pi$.

1. ____________
2. ____________
3. ____________
4. ____________
5. ____________
6. ____________
7. ____________
8. ____________
9. ____________
10. ____________

Name Score

**Write with a positive or zero exponent. Then evaluate.**

1. $6^{-2}$

2. $\frac{2^{-4}}{2^2}$

3. $\frac{5^{-2}}{5}$

1. ____________
2. ____________
3. ____________

**Simplify.**

4. $a^{-5}$

5. $xy^{-2}$

6. $x^3x^{-3}$

7. $\frac{a^{-7}}{a^7}$

8. $\frac{x^{-1}y^2}{x}$

9. $\left(x^2\right)^{-4}$

10. $\left(a^{-1}b^2\right)^0$

11. $\frac{6x^2}{2x}$

12. $\frac{18x^5}{-6x^2}$

13. $\frac{a^7b^9}{a^9b^3}$

14. $\frac{x^8y^6}{x^4y^3}$

15. $\frac{(3y)^2}{9y}$

16. $\frac{(5a)^3}{25a^2}$

17. $\frac{(4a)^3}{\left(-2a^2\right)^2}$

18. $\frac{\left(-4a^2\right)^2}{(2a)^2}$

19. $\frac{-60a^3b^7}{80ab^3}$

20. $\frac{15a^2b^5}{-90a^3b^3}$

21. $\frac{-6x^2y^5}{33y^3z^3}$

22. $\frac{-16x^7y^5}{-64x^3y^2}$

23. $\frac{-a^9b^3c^8}{a^6bc^7}$

24. $\frac{\left(-4x^2y\right)^2}{-12xy^4}$

4. ____________
5. ____________
6. ____________
7. ____________
8. ____________
9. ____________
10. ____________
11. ____________
12. ____________
13. ____________
14. ____________
15. ____________
16. ____________
17. ____________
18. ____________
19. ____________
20. ____________
21. ____________
22. ____________
23. ____________
24. ____________

Name Score

**Write in scientific notation.**

1. 0.0000356
2. 320,000
3. 0.00000037
4. 0.0000000051
5. 60,000,000,000
6. 46,000,000

1. ______
2. ______
3. ______
4. ______
5. ______
6. ______

**Write in decimal notation.**

7. $2.14\times10^{-6}$
8. $5.3\times10^{-9}$
9. $6.7\times10^{9}$
10. $5.23\times10^{6}$
11. $4.5\times10^{-3}$
12. $5.24\times10^{7}$

7. ______
8. ______
9. ______
10. ______
11. ______
12. ______

**Simplify.**

13. $(4\times10^{-10})(7\times10^{7})$
14. $(7.8\times10^{-4})(2.1\times10^{-5})$
15. $\dfrac{8\times10^{-2}}{2\times10^{7}}$
16. $\dfrac{2.8\times10^{5}}{4\times10^{-3}}$
17. $\dfrac{0.00072}{40,000,000}$

13. ______
14. ______
15. ______
16. ______
17. ______

**Solve. Write the answer in scientific notation.**

18. A high-speed centrifuge makes $8\times10^{7}$ revolutions each minute. Find the time in seconds for the centrifuge to make 1 revolution.
19. A space vehicle travels at an average velocity of $6.0\times10^{5}$ miles per day. How far does it travel each week?

18. ______
19. ______

Name Score

**Simplify.**

1. $\dfrac{4x+4}{4}$

2. $\dfrac{7y+7}{7}$

3. $\dfrac{8a-12}{4}$

4. $\dfrac{20a-30}{10}$

5. $\dfrac{12b-30}{6}$

6. $\dfrac{2a^2+3a}{a}$

7. $\dfrac{5y^2+3y}{y}$

8. $\dfrac{3b^3-2b}{b}$

9. $\dfrac{20y^2-10y}{10y}$

10. $\dfrac{x^3+2x^2-4x}{x}$

11. $\dfrac{x^4-2x^3-x^2}{x^2}$

12. $\dfrac{a^6-6a^5-4a^4}{a^3}$

13. $\dfrac{4x^2y^2-12xy}{4xy}$

14. $\dfrac{10y^6-25y^4}{-5y^4}$

15. $\dfrac{8x^4-6x^2}{-2x^2}$

16. $\dfrac{3y^4-12y^3}{-3y^3}$

17. $\dfrac{6x^2-9x^4}{-3x^2}$

18. $\dfrac{4x^2+x-5}{x}$

19. $\dfrac{6y^2+7y-5}{y}$

20. $\dfrac{-20x^2+5x-4}{x}$

21. $\dfrac{8a^2b-10ab+12ab^2}{2ab}$

22. $\dfrac{20a^2b+10ab-40ab^2}{10ab}$

23. $\dfrac{5x^2y+6xy-2xy^2}{xy}$

24. $\dfrac{6ab^2-24ab+18a^2b^2}{6ab}$

1. ______
2. ______
3. ______
4. ______
5. ______
6. ______
7. ______
8. ______
9. ______
10. ______
11. ______
12. ______
13. ______
14. ______
15. ______
16. ______
17. ______
18. ______
19. ______
20. ______
21. ______
22. ______
23. ______
24. ______

Name Score

**Simplify.**

1. $(x^2+6x+9)\div(x+3)$
2. $(x^2+8x+16)\div(x+4)$
3. $(x^2-12x+36)\div(x-6)$
4. $(a^2-14a+49)\div(a-7)$
5. $(x^2-2x-8)\div(x+2)$
6. $(y^2-y-30)\div(y-6)$
7. $(x^2-3x-18)\div(x-6)$
8. $(y^2+5y-14)\div(x+7)$
9. $(2x^2+5x+3)\div(x+1)$
10. $(2y^2-11y+14)\div(y-2)$
11. $(2y^2+y-15)\div(y+3)$
12. $(2y^2+5y-12)\div(y+4)$
13. $(3x^2-27)\div(3x-9)$
14. $(2y^2+11y-21)\div(2y-3)$
15. $(x^2-1)\div(x+1)$
16. $(x^2+9)\div(x+3)$
17. $(4x^2-10x)\div(2x-3)$
18. $(x^2+9)\div(x+3)$
19. $(a^2+3a+10)\div(a+5)$
20. $(b^2-7b-8)\div(b-4)$
21. $(10y^2-29y-21)\div(2y-7)$
22. $(12a^2-6a-6)\div(2a+3)$
23. $(6a^2+11a-10)\div(3a-2)$
24. $(15x^2+14x-8)\div(5x-2)$

1. ____________
2. ____________
3. ____________
4. ____________
5. ____________
6. ____________
7. ____________
8. ____________
9. ____________
10. ____________
11. ____________
12. ____________
13. ____________
14. ____________
15. ____________
16. ____________
17. ____________
18. ____________
19. ____________
20. ____________
21. ____________
22. ____________
23. ____________
24. ____________

Name Score

**Divide by using synthetic division.**

1. $(2x^2 - x - 10) \div (x + 2)$

2. $(4x^2 + 10x - 5) \div (x + 3)$

3. $(2x^2 - 17x + 35) \div (x - 5)$

4. $(3x^2 - 7x - 15) \div (x - 4)$

5. $(4x^2 - 3) \div (x - 1)$

6. $(4x^2 - 8) \div (x + 3)$

7. $(2x^3 + 3x^2 + 5x + 14) \div (x + 2)$

8. $(3x^3 + 2x^2 - 8x + 3) \div (x - 1)$

9. $(3x^3 + x - 15) \div (x - 3)$

10. $(x^3 - 4x^2 + 10) \div (x + 2)$

11. $(4x^3 + 15x^2 - 6x - 8) \div (x + 4)$

12. $(3x^3 + 8x^2 - 6x + 2) \div (x - 1)$

13. $(2x^3 - 5x^2 - 6x + 10) \div (x - 5)$

14. $(4x^3 + 9x^2 - 8x + 3) \div (x + 3)$

15. $\dfrac{3x^4 - 6x^3 + 4x^2 - 3x - 4}{x + 2}$

16. $\dfrac{3x^4 - 10x^3 - x^2 + 23x - 2}{x - 3}$

17. $\dfrac{3x^4 - x^2 + 3}{x - 2}$

18. $\dfrac{x^4 - 4x^2 - 18}{x - 3}$

1. ______________
2. ______________
3. ______________
4. ______________
5. ______________
6. ______________
7. ______________
8. ______________
9. ______________
10. ______________
11. ______________
12. ______________
13. ______________
14. ______________
15. ______________
16. ______________
17. ______________
18. ______________

Name Score

**Use the Remainder Theorem to evaluate the polynomial function.**

**1.** $P(x) = 3x^2 - 2x - 5;\ P(3)$

**2.** $P(x) = x^3 - 5x^2 + 2x + 7;\ P(-2)$

**3.** $R(t) = 4t^3 - t^2 + 5t - 11;\ R(-3)$

**4.** $F(x) = 3x^3 - x + 5;\ F(4)$

**5.** $S(t) = 5t^4 - t;\ S(-1)$

**6.** $R(t) = 4t^3 - 3t + 2t - 5;\ R(2)$

**7.** $P(x) = x^4 - 3x^3 + 7x - 5;\ P(5)$

**8.** $Q(x) = x^5 - 5x^3 - x + 6;\ Q(2)$

**9.** $P(t) = t^4 - 4t^3 + 5t^2 - t + 4;\ P(-4)$

**10.** $Q(y) = 4y^3 + 5y^2 - 7;\ Q(-2)$

**11.** $Z(y) = y^4 - y^3 + y^2 - y + 1;\ Z(-1)$

**12.** $R(x) = x^3 - 4x^2 + 6x - 5;\ R(3)$

**13.** $S(x) = x^3 - 3x^2 + 3x - 1;\ S(1)$

**14.** $R(z) = z^4 - 2z^2 + 4;\ R(-5)$

**15.** $F(x) = x^5 + x^4 - x^3 + 44x^2 - x + 3;\ F(-4)$

**16.** $Y(z) = z^3 + 4z^2 + 4z + 1;\ Y(-3)$

**1.** ____________

**2.** ____________

**3.** ____________

**4.** ____________

**5.** ____________

**6.** ____________

**7.** ____________

**8.** ____________

**9.** ____________

**10.** ____________

**11.** ____________

**12.** ____________

**13.** ____________

**14.** ____________

**15.** ____________

**16.** ____________

Name Score

**Factor.**

1. $4a+4$
2. $6c-6$
3. $8-4a^2$
4. $9+21x^2$
5. $3x+9$
6. $10a^2+15$
7. $24a-8$
8. $24x+12$
9. $9x-6$
10. $12xy-16y$
11. $6b^2-5b^3$
12. $20x^3-24x^2$
13. $12a^5-36a^2$
14. $24a^3b^4-18a^2b^2$
15. $4a^5b+6ab^4$
16. $25x^2y^2-15x^3y$
17. $3x^2y-5xy^2$
18. $8a^3b^2-12a^2b^3$
19. $x^3-5x^2+7x$
20. $4x^2-16x+20$
21. $6y^2-9y+12$
22. $b^4-3b^3+7b^2$
23. $4x^2-8x^3+12x^4$
24. $12y^2-16y+48$
25. $5y^4+10y^3-35y$
26. $4x^4+12x^3-28x^2$
27. $45a^4b^2-75a^3b+30ab^4$

1. ______
2. ______
3. ______
4. ______
5. ______
6. ______
7. ______
8. ______
9. ______
10. ______
11. ______
12. ______
13. ______
14. ______
15. ______
16. ______
17. ______
18. ______
19. ______
20. ______
21. ______
22. ______
23. ______
24. ______
25. ______
26. ______
27. ______

Name Score

**Factor.**

1. $x(a+b)+3(a+b)$
2. $a(x-y)+5(x-y)$
3. $x(b-1)-y(b-1)$
4. $y(a-1)-(a-1)$
5. $a(y+3)-(y+3)$
6. $3x(y-7)-(y-7)$
7. $2x(x-5)-(x-5)$
8. $5y(x-3)-(3-x)$
9. $x(a-2b)+y(a-2b)$
10. $a(x-9)-(x-9)$
11. $b(a-5)-2(a-5)$
12. $c(x-3y)+d(x-3y)$
13. $3x(a+1)+4(a+1)$
14. $ax-4x+4y-ay$
15. $ax-3bx+3by-ay$
16. $ax-9x-9+a$
17. $ax-7a+21-3x$
18. $2ax-8x-4y+ay$
19. $am-9m-9n+an$
20. $3mn-6m-2+n$
21. $2ab-2a-bc+c$
22. $ax-2x-3ay+6y$
23. $2a-2b+bc-ac$
24. $mn-m+2-2n$
25. $cx-7x-7y+cy$
26. $ab-ac-3c+3b$
27. $5x-5y-yz+xz$

1. ________
2. ________
3. ________
4. ________
5. ________
6. ________
7. ________
8. ________
9. ________
10. ________
11. ________
12. ________
13. ________
14. ________
15. ________
16. ________
17. ________
18. ________
19. ________
20. ________
21. ________
22. ________
23. ________
24. ________
25. ________
26. ________
27. ________

Name Score

**Factor.**

**1.** $a^2 - 2a - 35$

**2.** $a^2 - 4a + 3$

**3.** $a^2 + 3a - 10$

**4.** $b^2 - 7b + 10$

**5.** $b^2 + 8b + 15$

**6.** $b^2 + 6b - 16$

**7.** $y^2 - y - 30$

**8.** $y^2 + 5y - 66$

**9.** $x^2 - 4x - 60$

**10.** $y^2 - 7y + 10$

**11.** $y^2 - 9y + 18$

**12.** $x^2 - 12x + 36$

**13.** $a^2 + 3b - 28$

**14.** $x^2 + 10x + 16$

**15.** $b^2 - 11b - 180$

**16.** $x^2 + 10x + 25$

**17.** $b^2 + 10b + 16$

**18.** $x^2 - 7x - 60$

**19.** $x^2 + 10x - 56$

**20.** $x^2 - 8x - 128$

**21.** $x^2 - 4x - 77$

**22.** $b^2 - 20b + 84$

**23.** $b^2 - 21b + 108$

**24.** $b^2 - 27b + 180$

**25.** $a^2 + 16a + 63$

**26.** $x^2 - 19x + 60$

**27.** $x^2 - 25x + 84$

**1.** ______
**2.** ______
**3.** ______
**4.** ______
**5.** ______
**6.** ______
**7.** ______
**8.** ______
**9.** ______
**10.** ______
**11.** ______
**12.** ______
**13.** ______
**14.** ______
**15.** ______
**16.** ______
**17.** ______
**18.** ______
**19.** ______
**20.** ______
**21.** ______
**22.** ______
**23.** ______
**24.** ______
**25.** ______
**26.** ______
**27.** ______

Name Score

**Factor.**

1. $2x^2+8x+6$
2. $3x^2+15x+12$
3. $3a^2+3a-36$
4. $4x^2-8x-12$
5. $ab^2+5ab-14a$
6. $xy^2-7xy+10x$
7. $xy^2+11xy+28x$
8. $x^3-8x^2+12x$
9. $3y^3-18y^2+15y$
10. $4x^2+28x-32$
11. $2x^2-2x-12$
12. $3z^2-21z-24$
13. $z^3+14z^2+45z$
14. $c^3+17c^2+30c$
15. $x^2-9xy+14y^2$
16. $4y^2-16-128$
17. $3x^2+3x-18$
18. $2x^2-14x-60$
19. $3y^2-24y+45$
20. $5y^2-50y+45$
21. $2x^3+2x^2-12x$
22. $y^2+7xy-8x^2$
23. $x^3+3x^2-40x$
24. $10x^3+60x^2+80x$
25. $x^4+2x^2-63$
26. $2x^2+14x-240$
27. $a^4+a^3-72a^2$

1. ______
2. ______
3. ______
4. ______
5. ______
6. ______
7. ______
8. ______
9. ______
10. ______
11. ______
12. ______
13. ______
14. ______
15. ______
16. ______
17. ______
18. ______
19. ______
20. ______
21. ______
22. ______
23. ______
24. ______
25. ______
26. ______
27. ______

Name Score

**Factor.**

1. $2x^2 - 5x + 2$
2. $3x^2 - 2x - 1$
3. $3x^2 + x - 2$
4. $2b^2 - 13b + 6$
5. $3a^2 - 7a + 2$
6. $4x^2 + 4x - 3$
7. $5a^2 + 13a - 6$
8. $6y^2 + 5y - 6$
9. $6x^2 + x - 5$
10. $5x^2 - 3x - 2$
11. $7y^2 + 8y + 1$
12. $14x^2 + 9x + 1$
13. $7y^2 + 18y + 8$
14. $9a^2 - 3a - 2$
15. $3a^2 - 5a - 12$
16. $3x^2 - 10x - 8$
17. $6x^2 - 5x - 6$
18. $4y^2 + 25y + 25$
19. $7x^2 + 20x - 3$
20. $5x^2 + 2x - 7$
21. $10x^2 - 11x - 6$
22. $8x^2 - 26x + 15$
23. $12x^2 - 7x - 10$
24. $8x^2 - 2x - 15$
25. $10x^2 - 21x - 10$
26. $15x^2 - 26x + 8$
27. $12y^2 - 16y - 3$

1. ______
2. ______
3. ______
4. ______
5. ______
6. ______
7. ______
8. ______
9. ______
10. ______
11. ______
12. ______
13. ______
14. ______
15. ______
16. ______
17. ______
18. ______
19. ______
20. ______
21. ______
22. ______
23. ______
24. ______
25. ______
26. ______
27. ______

Name Score

**Factor.**

1. $3x^2 - 6x + 3$
2. $8x^2 + 22x - 6$
3. $6y^2 - 20y + 14$
4. $15y^2 - 5y - 10$
5. $2a^2b + 8ab - 24b$
6. $4x^2 - 25xy + 6y^2$
7. $x^2 - 11xy + 30y^2$
8. $2x^3 - 7x^2 - 4x$
9. $15 + 8x + x^2$
10. $21 + 4x - x^2$
11. $32 + 4x - x^2$
12. $4y^2 - 4y - 3$
13. $12y^2 + 12y - 9$
14. $12x^2 - 19x + 4$
15. $5x^2 - 35x - 40$
16. $5x^3 - 6x^2 + x$
17. $3x^2 - 11x - 20$
18. $12y^2 - 12y - 9$
19. $8x^2 + 8x - 6$
20. $10a^3 - 15a^2 - 25a$
21. $6x^3 + 17x^2 + 5x$
22. $6x^3 - 26x^2 + 8x$
23. $10b^2 + 14b - 12$
24. $22z - 9z^2 - z^3$
25. $6x^3 - 32x^2 + 10x$
26. $19 + 18x - x^2$
27. $13x^2 + 49x - 12$

1. ____________
2. ____________
3. ____________
4. ____________
5. ____________
6. ____________
7. ____________
8. ____________
9. ____________
10. ____________
11. ____________
12. ____________
13. ____________
14. ____________
15. ____________
16. ____________
17. ____________
18. ____________
19. ____________
20. ____________
21. ____________
22. ____________
23. ____________
24. ____________
25. ____________
26. ____________
27. ____________

Name Score

**Factor.**

1. $x^2 - 9$
2. $x^2 - 36$
3. $9x^2 - 1$
4. $9x^2 - 16$
5. $36x^2 - 64$
6. $a^2b^2 - 121$
7. $x^2 + 9$
8. $a^2 + 144$
9. $16 - a^2b^2$
10. $49 - x^2y^2$
11. $36a^2 - 25b^4$
12. $a^{2n} - 9$
13. $b^{2n} - 36$
14. $a^2 + 18a + 81$
15. $9a^2 + 3a - 1$
16. $4x^2 + 8x - 3$
17. $b^2 + 9b + 9$
18. $y^2 - 3y + 24$
19. $x^2 + 4xy + 4y^2$
20. $4x^2y^2 + 29xy + 25$
21. $25a^2 - 20ab + 4b^2$
22. $4a^2 - 20ab + 25b^2$
23. $x^{2n} + 8x^n + 16$
24. $y^{2n} - 20y^n + 100$

1. __________
2. __________
3. __________
4. __________
5. __________
6. __________
7. __________
8. __________
9. __________
10. __________
11. __________
12. __________
13. __________
14. __________
15. __________
16. __________
17. __________
18. __________
19. __________
20. __________
21. __________
22. __________
23. __________
24. __________

Name Score

**Factor.**

1. $x^3 - 8$
2. $x^3 - 64$
3. $27x^3 - 1$
4. $8x^3 + 27$
5. $x^3 + 8y^3$
6. $x^3 + 216y^3$
7. $m^3 - n^3$
8. $125x^3 - y^3$
9. $216a^3 + 1$
10. $1 - 216a^3$
11. $64x^3 - 27y^3$
12. $27x^3 + 8y^3$
13. $x^3y^3 + 8$
14. $27x^3y^3 + 64$
15. $25x^3 - y^3$
16. $8x^3 + 9y^2$
17. $16a^3 - 27$
18. $27x^3 - 125$
19. $125 - x^3$
20. $64 + x^3$
21. $8x^3 - 25$
22. $(a+b)^3 - 8$
23. $a^3 - (a+b)^3$
24. $16x^3 - 25$
25. $x^{3n} - y^{3n}$
26. $y^{3n} + 64$
27. $a^{3n} - 125$

1. ______
2. ______
3. ______
4. ______
5. ______
6. ______
7. ______
8. ______
9. ______
10. ______
11. ______
12. ______
13. ______
14. ______
15. ______
16. ______
17. ______
18. ______
19. ______
20. ______
21. ______
22. ______
23. ______
24. ______
25. ______
26. ______
27. ______

Name Score

**Factor.**

1. $x^2y^2 - 8xy - 20$

2. $x^2y^2 - 13xy + 40$

3. $a^2b^2 + 11ab + 28$

4. $a^2b^2 - 15ab + 56$

5. $x^4 - 10x^2 + 16$

6. $y^4 - 11y^2 - 42$

7. $b^4 - 10b^2 + 21$

8. $b^4 + 12b^2 + 27$

9. $a^4 - 7a^2 - 60$

10. $a^4b^4 + 10a^2b^2 - 24$

11. $a^4b^4 + 9a^2b^2 - 33$

12. $a^{2n} - 2a^n - 15$

13. $2x^2y^2 + 3xy - 20$

14. $3x^2y^2 + 2xy - 8$

15. $3a^2b^2 - 17ab + 10$

16. $2a^{2n} + 9x^n + 9$

17. $2a^{2n} - 9x^n - 5$

18. $6a^{2n} + 7a^n + 2$

1. ______
2. ______
3. ______
4. ______
5. ______
6. ______
7. ______
8. ______
9. ______
10. ______
11. ______
12. ______
13. ______
14. ______
15. ______
16. ______
17. ______
18. ______

Name Score

**Factor.**

1. $3x^2 - 12$
2. $2x^2 - 50$
3. $x^3 + 2x^2 + x$
4. $y^3 + 8y^2 + 16y$
5. $x^4 + x^3 - 6x^2$
6. $a^4 - 3a^3 + 40a^2$
7. $3b^2 + 30b + 63$
8. $4y^2 - 32y + 28$
9. $2a^2 - 18a - 44$
10. $x^3 - 8x^2 - 20x$
11. $b^3 - 5b^2 - 6b$
12. $5a^3 - 30a^2 + 45a$
13. $4x^2 - 6x + 2$
14. $21x^4 - 11x^3 + 5x^2$
15. $x^4 - 16x^2$
16. $15x^3 - 18x^2 + 3x$
17. $9a^3 + 39a^2 + 12a$
18. $3xy^2 + 11xy - 20x$
19. $24 + 6x - 3x^2$
20. $a^2b^2 + 7ab^2 - 8b^2$
21. $4x^2y + 12xy + 8y$
22. $18a^3 + 54a^2 + 36a$
23. $x^2y^2 - y^2$
24. $5x^2 - 45y^2$
25. $x^4 - 9x^2$
26. $a(3x - 3) + b(3x - 3)$
27. $y^2(a + b) - 4(a + b)$

1. ______
2. ______
3. ______
4. ______
5. ______
6. ______
7. ______
8. ______
9. ______
10. ______
11. ______
12. ______
13. ______
14. ______
15. ______
16. ______
17. ______
18. ______
19. ______
20. ______
21. ______
22. ______
23. ______
24. ______
25. ______
26. ______
27. ______

Name Score

**Factor.**

1. $4x^2-8x+4$
2. $5x^2-20$
3. $6x^2-6$
4. $16x^2-36$
5. $x^4-81$
6. $81ax^4-1$
7. $12x^4-75x^2$
8. $3a-3a^4$
9. $a^6b^3+27a^3$
10. $3x^4+9x^3-120x^2$
11. $x^4-3x^3-28x^2$
12. $a^4+5a^2-36$
13. $a^4-8a^2-9$
14. $3a^4-24a^3+48a^2$
15. $a^4-5a^2+4$
16. $x^4-16$
17. $10x^2-5x-15$
18. $6y^2+30y+36$
19. $15x^3y-18x^2y^2-24xy^3$
20. $16x^4+4x^3y-30x^2y^2$
21. $a^{2n+3}-8a^{n+2}+16a^2$
22. $x^{2n+1}+4x^{n+1}+4x$
23. $2x^{n-2}-7x^{n+1}+6x^n$
24. $2a^{n+2}+5a^{n+1}-3a^n$

1. ______
2. ______
3. ______
4. ______
5. ______
6. ______
7. ______
8. ______
9. ______
10. ______
11. ______
12. ______
13. ______
14. ______
15. ______
16. ______
17. ______
18. ______
19. ______
20. ______
21. ______
22. ______
23. ______
24. ______

Name Score

**Solve.**

**1.** $(y+1)(y+2)=0$

**2.** $(y-4)(y-6)=0$

**3.** $(z-6)(z-1)=0$

**4.** $(x+7)(x-5)=0$

**5.** $x(x-8)=0$

**6.** $x(x+1)=0$

**7.** $a(a-4)=0$

**8.** $y(3y+2)=0$

**9.** $t(2t-5)=0$

**10.** $3a(2a-1)=0$

**11.** $2b(4b+3)=0$

**12.** $(b+1)(b-4)=0$

**13.** $(b-7)(b+4)=0$

**14.** $x^2-4x-21=0$

**15.** $x^2+6x-16=0$

**16.** $x^2-5x=6$

**17.** $x^2-5x=14$

**18.** $2a^2-a=3$

**19.** $4t^2-13=-3$

**20.** $5a^2+13a=6$

**21.** $2x^2+5x=-2$

**22.** $x(x+10)=11$

**23.** $y(y-9)=-18$

**24.** $x(x+5)=50$

**25.** $(2x+3)(x-1)=25$

**26.** $(z+1)(z-9)=39$

**27.** $(y+3)(2y-1)=9$

**1.** ______
**2.** ______
**3.** ______
**4.** ______
**5.** ______
**6.** ______
**7.** ______
**8.** ______
**9.** ______
**10.** ______
**11.** ______
**12.** ______
**13.** ______
**14.** ______
**15.** ______
**16.** ______
**17.** ______
**18.** ______
**19.** ______
**20.** ______
**21.** ______
**22.** ______
**23.** ______
**24.** ______
**25.** ______
**26.** ______
**27.** ______

Name ____________ Score ____________

**Solve.**

**1.** The square of a positive number is four more than three times the positive number. Find the number.

**2.** The square of a negative number is forty-eight more than twice the negative number. Find the number.

**3.** The sum of two numbers is seven. The sum of the squares of the two numbers is twenty-nine. Find the two numbers.

**4.** The sum of two numbers is ten. The sum of the squares of the two numbers is fifty-two. Find the two numbers.

**5.** The sum of two numbers is twelve. The product of the two numbers is twenty-seven. Find the two numbers.

**6.** The square of the sum of a number and six is eighty-one. Find the number.

**7.** The sum of the squares of two consecutive positive integers is sixty-one. Find the two integers.

**8.** The product of two consecutive even positive integers is one hundred twenty. Find the integers.

**9.** The length of the base of a triangle is three times the height. The area of the triangle is 96 $ft^2$. Find the base and height of the triangle.

**10.** The length of a rectangle is four times the width. The area is 144 $in^2$. Find the length and width.

**11.** The width of a rectangle is 7 ft less than the length. The area of the rectangle is 120 $ft^2$. Find the length and width of the rectangle.

**12.** The length of each side of a square is extended by 3 in. The area of the resulting square is 121 $in^2$. Find the length of a side of the original square.

**1.** ____________

**2.** ____________

**3.** ____________

**4.** ____________

**5.** ____________

**6.** ____________

**7.** ____________

**8.** ____________

**9.** ____________

**10.** ____________

**11.** ____________

**12.** ____________

Name Score

**Evaluate.**

1. Given $f(x)=\dfrac{-7x}{x^2-3x+6}$, find $f(2)$.

2. Given $g(x)=\dfrac{x+8}{9x^2-6x+3}$, find $g(-2)$.

3. Given $f(x)=\dfrac{x-8}{3x^2-2x-11}$, find $f(-3)$.

4. Given $g(x)=\dfrac{-6}{x^2+5x-9}$, find $g(4)$.

5. Given $f(x)=\dfrac{x+14}{2x^2-7x+2}$, find $f(-5)$.

6. Given $g(x)=\dfrac{-12}{x^2-5x-4}$, find $g(3)$.

7. Given $f(x)=\dfrac{5}{x^2-9x+1}$, find $f(4)$.

8. Given $g(x)=\dfrac{x+3}{3x^2+4x-5}$, find $g(1)$.

**Find the domain.**

9. $f(x)=\dfrac{x-7}{2x+6}$

10. $g(x)=\dfrac{x+11}{x^2-x-12}$

11. $f(x)=\dfrac{6x+5}{x^2+2x+1}$

12. $g(x)=\dfrac{4x-3}{5x-35}$

13. $f(x)=\dfrac{6}{x^2-9}$

14. $g(x)=\dfrac{5x+3}{x^2-3x-10}$

15. $f(x)=\dfrac{-41}{6x+24}$

16. $g(x)=\dfrac{14x-3}{x^2-x-6}$

1. ______
2. ______
3. ______
4. ______
5. ______
6. ______
7. ______
8. ______
9. ______
10. ______
11. ______
12. ______
13. ______
14. ______
15. ______
16. ______

Name Score

**Simplify.**

1. $\frac{3+9x}{3}$

2. $\frac{12x^2-4x}{4x}$

3. $\frac{10x^2(x+2)}{2x(x+2)}$

4. $\frac{20x^4(x+7)}{15x^3(x+7)}$

5. $\frac{3x-9}{3x-x^2}$

6. $\frac{12x^3-6x^2+18x}{6x}$

7. $\frac{4x^{3n}-12x^{2n}}{16x^{2n}}$

8. $\frac{a^2-8a+15}{a^2+2a-35}$

9. $\frac{x^2-5x}{x^2+5x}$

10. $\frac{x^{3n}+x^{2n}y^n}{x^{3n}-x^{2n}y^n}$

11. $\frac{9a^n}{3a^{2n}-9a^n}$

12. $\frac{12-x-x^2}{2x^2-7x+3}$

13. $\frac{3x^2-13x-10}{10-17x+3x^2}$

14. $\frac{x^2-3xy-4y^2}{x^2+4xy+3y^2}$

15. $\frac{x^2+7x-18}{x^2-5x+6}$

16. $\frac{27x^3-y^3}{9x^2-y^2}$

17. $\frac{6x^2y^2+7xy^2+2}{6x^2y^2+19xy+10}$

18. $\frac{x^2-9}{a(x+3)-b(x+3)}$

19. $\frac{x^4-7x^2-8}{x^4+9x^2+8}$

20. $\frac{x^4-16}{x^4-x^2-12}$

21. $\frac{x^2y^2+8xy-36}{x^2y^2+2xy-63}$

22. $\frac{a^{2n}-3a^n-28}{a^{2n}-a^n-42}$

23. $\frac{a^{2n}-a^n-6}{a^{2n}+2a^n-15}$

1. __________
2. __________
3. __________
4. __________
5. __________
6. __________
7. __________
8. __________
9. __________
10. __________
11. __________
12. __________
13. __________
14. __________
15. __________
16. __________
17. __________
18. __________
19. __________
20. __________
21. __________
22. __________
23. __________

Name Score

**Multiply or divide.**

1. $\dfrac{30a^3b^5}{25x^3y^3} \cdot \dfrac{5x^2y^3}{6a^2b^3}$

2. $\dfrac{x^3y^2}{x^2-6x-7} \cdot \dfrac{2x^2-15x+7}{x^3y^4}$

3. $\dfrac{x^2+2x-15}{x^2+3x-10} \cdot \dfrac{x^2-x-2}{x^2-1}$

4. $\dfrac{x^2-x-6}{15-8x+x^2} \cdot \dfrac{x^2-x-20}{x^2+4x+4}$

5. $\dfrac{x^2+4x-12}{x^2-7x+6} \cdot \dfrac{x^2-3x+2}{x^2-4x+4}$

6. $\dfrac{9x^2+3x-2}{6x^2+13x-5} \cdot \dfrac{2x^2-13x+20}{8+10x-3x^2}$

7. $\dfrac{x^{n+1}+3x^n}{6x^2-9x} \cdot \dfrac{8x-12}{x^{n+1}-x^n}$

8. $\dfrac{x^3-8}{2x^2-7x+6} \cdot \dfrac{2x^2+x-6}{x^2+2x+4}$

9. $\dfrac{x^2-6x+5}{3x^2-5x+2} \cdot \dfrac{3x^2-8x+4}{x^2-25}$

10. $\dfrac{6x^2y^4}{15a^4b^3} \div \dfrac{16x^4y^3}{5a^3b^3}$

11. $\dfrac{4x-12}{10x^2+15x} \div \dfrac{6x^3-18x^2}{14x^4+21x^3}$

12. $\dfrac{x^4y^3}{x^2+4x-5} \div \dfrac{x^2y^2}{2x^2+11x+5}$

13. $\dfrac{x^2-4x+3}{x^2+5x-6} \div \dfrac{x^2-2x-3}{x^2-x-2}$

14. $\dfrac{x^{n+1}-2x^n}{3x^2-6x} \div \dfrac{x^{n+1}+2x^n}{6x^2+12x}$

15. $\dfrac{3x^2-11x+10}{3x^2+x-10} \div \dfrac{2x^2-15x+28}{2x^2-3x-14}$

16. $\dfrac{2x^2+x-15}{2x^2-9x+10} \div \dfrac{3x^2+5x-12}{6x^2+10x-24}$

17. $\dfrac{4x^2-25}{2x^2-3x-20} \div \dfrac{6x^2+3x-45}{6x^2-6x-72}$

18. $\dfrac{x^{2n}+5x^n+6}{x^{2n}-x^n-6} \div \dfrac{x^{2n}+x^n-6}{x^{2n}+6x^n+9}$

1. ______
2. ______
3. ______
4. ______
5. ______
6. ______
7. ______
8. ______
9. ______
10. ______
11. ______
12. ______
13. ______
14. ______
15. ______
16. ______
17. ______
18. ______

Name Score

**Simplify.**

1. $\dfrac{2}{3xy}-\dfrac{5}{3xy}+\dfrac{3}{3xy}$

2. $-\dfrac{4}{5x^2}+\dfrac{9}{5x^2}-\dfrac{4}{5x^2}$

3. $\dfrac{x}{x^2-4x-5}-\dfrac{5}{x^4-4x-5}$

4. $\dfrac{3x}{3x^2+13x-10}-\dfrac{2}{3x^2+13x-10}$

5. $\dfrac{x}{x^2-y^2}-\dfrac{y}{x^2-y^2}$

6. $\dfrac{4x}{4x^2+3x-10}-\dfrac{5}{4x^2+3x-10}$

7. $\dfrac{3}{5xy^2}-\dfrac{4}{15y}-\dfrac{7}{3xy}$

8. $\dfrac{3}{4a^2b}-\dfrac{5}{8ab^2}-\dfrac{1}{12a^2b^2}$

9. $\dfrac{3x-2}{15x}-\dfrac{2x-5}{10x}$

10. $\dfrac{5a}{a-2}-4+\dfrac{2}{a}$

11. $\dfrac{3x-4}{x+5}-\dfrac{x^2-8x-20}{x^2+9x+20}$

12. $\dfrac{2}{x+5}-\dfrac{4x}{x^2+10x+25}$

13. $\dfrac{x}{x+5}-\dfrac{5-x}{x^2-25}$

14. $\dfrac{1}{x+4}-\dfrac{3x}{x^2+8x+16}$

15. $\dfrac{5x}{x-4}-\dfrac{4x}{x-5}$

16. $\dfrac{6a}{a+4}-\dfrac{11a}{a-1}$

17. $\dfrac{x-2}{x+4}-\dfrac{x^2-3x-18}{x^2+7x+12}$

18. $\dfrac{3x^n-9}{x^{2n}+x^n-12}-\dfrac{x^n}{x^n-3}$

19. $\dfrac{3x-2}{4x^2-9}-\dfrac{4}{2x-3}$

20. $\dfrac{x^2+3}{5x^2-20}-\dfrac{8}{x+2}$

1. __________
2. __________
3. __________
4. __________
5. __________
6. __________
7. __________
8. __________
9. __________
10. __________
11. __________
12. __________
13. __________
14. __________
15. __________
16. __________
17. __________
18. __________
19. __________
20. __________

Name Score

**Simplify.**

1. $\dfrac{3+\dfrac{1}{x}}{9-\dfrac{1}{x^2}}$

2. $\dfrac{\dfrac{16}{x^2}-1}{1+\dfrac{4}{x}}$

3. $\dfrac{a-5}{\dfrac{25}{a}-a}$

4. $\dfrac{\dfrac{36}{a}-a}{6+a}$

5. $\dfrac{\dfrac{3}{a^2}+\dfrac{3}{a}}{\dfrac{3}{a^2}+\dfrac{3}{a}}$

6. $\dfrac{\dfrac{1}{x}+\dfrac{1}{4}}{\dfrac{16}{x^2}-1}$

7. $\dfrac{1+\dfrac{1}{x}-\dfrac{12}{x^2}}{1+\dfrac{3}{x}-\dfrac{18}{x^2}}$

8. $\dfrac{\dfrac{12}{x^2}-\dfrac{5}{x}-2}{\dfrac{8}{x^2}-\dfrac{10}{x}+3}$

9. $\dfrac{6+\dfrac{13}{x}-\dfrac{5}{x^2}}{2-\dfrac{11}{x}+\dfrac{15}{x^2}}$

10. $\dfrac{2+\dfrac{18}{x}+\dfrac{16}{x^2}}{5+\dfrac{2}{x}-\dfrac{3}{x^2}}$

11. $\dfrac{1-\dfrac{5}{2x-3}}{x-\dfrac{24}{2x-3}}$

12. $\dfrac{1+\dfrac{4}{3x+5}}{x-\dfrac{12}{3x+5}}$

13. $\dfrac{x+2-\dfrac{18}{x-1}}{x+12+\dfrac{42}{x-1}}$

14. $\dfrac{x+2-\dfrac{15}{x+4}}{x+5-\dfrac{6}{x+4}}$

15. $\dfrac{\dfrac{2}{a}-\dfrac{4}{a-3}}{\dfrac{1}{a}+\dfrac{3}{a-3}}$

16. $\dfrac{\dfrac{3}{a}-\dfrac{6}{a+4}}{\dfrac{4}{a}+\dfrac{4}{a+4}}$

17. $\dfrac{\dfrac{x-2}{x+2}-\dfrac{x+2}{x-2}}{\dfrac{x-2}{x+2}+\dfrac{x+2}{x-2}}$

18. $\dfrac{\dfrac{x}{x+4}-\dfrac{x}{x-4}}{\dfrac{x}{x+4}+\dfrac{x}{x-4}}$

1. ____________
2. ____________
3. ____________
4. ____________
5. ____________
6. ____________
7. ____________
8. ____________
9. ____________
10. ____________
11. ____________
12. ____________
13. ____________
14. ____________
15. ____________
16. ____________
17. ____________
18. ____________

Name Score

**Simplify.**

1. $\frac{x}{3}+\frac{3}{4}=\frac{x}{12}$

2. $\frac{x}{5}-\frac{3}{8}=\frac{x}{20}$

3. $\frac{6}{5x-3}=3$

4. $-2=\frac{10}{3x+4}$

5. $3-\frac{4}{x}=5$

6. $5-\frac{8}{x}=7$

7. $\frac{1}{x-2}=\frac{3}{x}$

8. $\frac{8}{x}=\frac{4}{x+1}$

9. $\frac{8}{x-3}=\frac{5}{x}$

10. $\frac{4}{x}=\frac{2}{x-6}$

11. $6+\frac{9}{a-4}=\frac{3a}{a-4}$

12. $\frac{-6}{a-6}=4-\frac{a}{a-6}$

13. $-\frac{4}{x+8}+2=\frac{6}{x+8}$

14. $7-\frac{3}{2x-3}=\frac{4}{2x-3}$

15. $\frac{3}{x^2-25}+\frac{1}{x+5}=\frac{8}{x-5}$

16. $\frac{6}{x-4}-\frac{3}{x+4}=\frac{9}{x^2-16}$

17. $\frac{11}{x^2-5x+6}=\frac{3}{x-2}+\frac{4}{x-3}$

18. $\frac{4}{x^2+8x+12}=\frac{7}{x+2}-\frac{3}{x+6}$

1. ____________
2. ____________
3. ____________
4. ____________
5. ____________
6. ____________
7. ____________
8. ____________
9. ____________
10. ____________
11. ____________
12. ____________
13. ____________
14. ____________
15. ____________
16. ____________
17. ____________
18. ____________

Name Score

**Solve.**

**1.** One printer can print the paychecks for the employees of a company in 60 min. A second printer can print the checks in 90 min. How long would it take to print the checks with both printers operating?

**2.** One solar heating panel can raise the temperature of water 1° in 30 min. A second solar heating panel can raise the temperature 1° in 45 min. How long would it take to raise the temperature of the water 1° with both solar panels operating?

**3.** A new printer can print checks three times faster than an old printer. The older printer can print the checks in 48 min. How long would it take to print the checks with both printers operating?

**4.** An experienced electrician can wire a room twice as fast as an apprentice electrician. Working together, the electricians can wire a room in 4 h. How long would it take the apprentice working alone to wire a room?

**5.** One member of a gardening team can mow and clean up a lawn in 12 h. With both members of the team working, the job can be done in 8 h. How long would it take the second member of the team, working alone to do the job?

**6.** A welder requires 18 h to do a job. After the welder and an apprentice work on a job for 6 h, the welder moves to another job. The apprentice finishes the job in 14 h. How long would it take the apprentice, working alone, to do the job?

**7.** Three computers can print out a task in 12 min, 18 min, and 36 min, respectively. How long would it take to complete the task with all three computers working?

**8.** Three machines are filling soda bottles. The machines can fill the daily quota of soda bottles in 10 h, 15 h, and 30 h, respectively. How long would it take to fill the daily quota of soda bottles with all three machines working?

**9.** The inlet pipe can fill a water tank in 48 min. The outlet pipe can empty the tank in 16 min. How long does it take to empty a full tank when both pipes are open?

**10.** An oil tank has two inlet pipes and one outlet pipe. One inlet pipe can fill the tank in 6 h, and the other inlet pipe can fill the tank in 12 h. The outlet pipe can empty the tank in 24 h. How long would it take to fill the tank with all three pipes open?

**1.** ______________

**2.** ______________

**3.** ______________

**4.** ______________

**5.** ______________

**6.** ______________

**7.** ______________

**8.** ______________

**9.** ______________

**10.** ______________

Name Score

**Solve.**

**1.** An express bus travels 455 mi in the same amount of time as a car travels 406 mi. The rate of the car is 7 mph less than the rate of the bus. Find the rate of the car.

**2.** A commercial jet travels 2700 mi in the same amount of time as a corporate jet travels 2175 miles. The rate of the commercial jet is 105 mph faster than the rate of the corporate jet. Find the rate of each jet.

**1.** ____________

**2.** ____________

**3.** A motorcycle travels 232 mi in the same amount of time as a car travels 200 mi. The rate the motorcycle is 8 mph faster than the rate of the car. Find the rate of the car.

**4.** A sales executive traveled 75 mi by car and then an additional 1400 mi by plane. The rate of the plane was eight times faster than the rate of the car. The total time for the trip was 5 h. Find the rate of the plane.

**3.** ____________

**4.** ____________

**5.** An express train and a car leave a town at 2 P.M. and head for a town 300 mi away. The rate of the express is one and one-half times the rate of the car. The train arrives 2 h ahead of the car. Find the rate of the car.

**6.** A corporate jet and a car start from a town at 7 A.M. and head for a town 660 mi away. The rate of the plane is six times the rate of the car. The plane arrives 10 h ahead of the car. Find the rate of the car.

**5.** ____________

**6.** ____________

**7.** A plane can fly at a rate of 212 mph in calm air. Traveling with the wind, the plane flew 720 mi in the same amount of time as it flew 552 mi against the wind. Find the rate of the wind.

**8.** A jet can fly at a rate of 500 mph in calm air. Traveling with the wind, the plane flew 2240 mi in the same amount of time as it flew 1760 mi against the wind. Find the rate of the wind.

**7.** ____________

**8.** ____________

**9.** A passenger train travels 165 mi in the same amount of time as a freight train travel 120 mi. The rate of the passenger train is 15 mph faster than the rate of the freight train. Find the rate of each train.

**10.** A plane can fly at a rate of 240 mph in calm air. Traveling with the wind, the plane flew 900 mi in the same amount of time as it flew 540 mi against the wind. Find the rate of the wind.

**9.** ____________

**10.** ____________

Name Score

**Simplify.**

1. $\frac{x+3}{18}=\frac{1}{2}$

2. $\frac{x+5}{12}=\frac{3}{4}$

3. $\frac{5}{x+3}=\frac{1}{2}$

4. $\frac{x}{2}=\frac{x-3}{4}$

5. $\frac{-2}{x-3}=\frac{5}{x}$

6. $\frac{5}{2}=\frac{20}{x+2}$

7. $\frac{6}{x+1}=\frac{4}{x-1}$

8. $\frac{10}{x-2}=\frac{15}{x-1}$

1. ______
2. ______
3. ______
4. ______
5. ______
6. ______
7. ______
8. ______

**Solve.**

9. The real estate tax for a house that cost \$75,000 is \$1600. At this rate, what is the value of a house for which the real estate tax is \$1920?

10. The license fee for a car that cost \$8000 is \$96. At the same rate, what is the license fee for a car that cost \$9600?

11. The scale on an architectural drawing is $\frac{1}{8}$ in. representing 1 ft. Find the dimensions of a room that measures $9\frac{1}{2}$ in. by 12 in. on the drawing.

12. Two pounds of pecans cost \$7.50. At this rate, how much would 25 lb of pecans cost? Round to the nearest cent.

13. One and one-fourth ounces of a medication are required for a 120-pound adult. At the same rate, how many additional ounces of medication are required for a 180-pound adult?

14. Ten ounces of an insecticide is mixed with 32 gal of water to make a spray for spraying a cotton field. How much additional insecticide is required to be mixed with 80 gal of water?

9. ______
10. ______
11. ______
12. ______
13. ______
14. ______

Name Score

**Solve.**

**1.** The distance ($d$) a spring will stretch varies directly as the force ($f$) applied to the spring. If a force of 12 lb is required to stretch a spring 8 in., what force is required to stretch the spring 12 in.?

**2.** The pressure ($p$) on a diver in the water varies directly as the depth ($d$). If the pressure is 8 lb/in$^2$ when the depth is 16 ft, what is the pressure when the depth is 20 ft?

**1.** ______________

**2.** ______________

**3.** The stopping distance ($s$) of a car varies directly as the square of its speed ($v$). If a car traveling 40 mph requires 150 ft to stop, find the stopping distance for a car traveling 50 mph.

**4.** The distance ($s$) a ball will roll down an inclined plane is directly proportional to the square of the time ($t$). If a ball rolls 8 ft in 1 s, how far will it roll in 4 s?

**3.** ______________

**4.** ______________

**5.** The speed ($v$) of a gear varies inversely as the number of teeth ($t$). If a gear that has 30 teeth makes 25 revolutions per minute, how many revolutions per minute will a gear that has 50 teeth make?

**6.** For a constant temperature, the pressure ($P$) of a gas varies inversely as the volume ($V$). If the pressure is 70 lb/in$^2$ when the volume is 210 ft$^3$, find the pressure when the volume is 140 ft$^3$.

**5.** ______________

**6.** ______________

**7.** The intensity ($I$) of a light source is inversely proportional to the square of the distance ($d$) from the source. If the intensity is 48 lumens at a distance of 12 ft, what is the intensity at a distance of 6 ft?

**8.** The length ($L$) of a rectangle of fixed area varies inversely as the width ($W$). If the length of a rectangle is 10 ft when the width is 8 ft, find the length of the rectangle when the width is 5 ft.

**7.** ______________

**8.** ______________

**9.** The frequency of vibration ($f$) in an open pipe organ varies inversely as the length ($L$) of the pipe. If the air in a pipe 3 m long vibrates 90 times per minute, find the frequency in a pipe that is 2 m long.

**10.** The current ($I$) in a wire varies directly as the voltage ($v$) and inversely as the resistance ($r$). If the current is 32 amps when the voltage is 192 volts and the resistance is 6 ohms, find the current when the voltage is 210 volts and the resistance is 12 ohms.

**9.** ______________

**10.** ______________

Name Score

**Solve the formula for the given variable.**

1. $a_n = a_1 + (n-1)d;\ n$

2. $A = \frac{1}{2}bh;\ b$

3. $x = -\frac{b}{2a};\ a$

4. $PV = nRT;\ n$

5. $F = \frac{Gm_1m_2}{r^2};\ G$

6. $\frac{P_1V_1}{T_1} = \frac{P_2V_2}{T_2};\ V_2$

7. $A = \frac{1}{2}h\left(b_1 + b_2\right);\ b_1$

8. $d = rt;\ r$

9. $I = \frac{100M}{C};\ C$

10. $E = aI(T-t);\ T$

11. $\frac{1}{t} = \frac{1}{a} + \frac{1}{b};\ a$

12. $d = \frac{a}{2}(2t-v);\ v$

13. $h = vt + \frac{1}{2}gt^2;\ g$

14. $R = \frac{C-S}{t};\ C$

15. $R = \frac{MB-T}{K};\ M$

16. $K = mr - mv;\ m$

17. $S = \pi r(r-I);\ I$

18. $A = \frac{P}{1-c};\ c$

1. ______________
2. ______________
3. ______________
4. ______________
5. ______________
6. ______________
7. ______________
8. ______________
9. ______________
10. ______________
11. ______________
12. ______________
13. ______________
14. ______________
15. ______________
16. ______________
17. ______________
18. ______________

Name Score

**Add.**

1. $25^{1/2}$

2. $27^{1/3}$

3. $16^{3/2}$

4. $16^{-1/2}$

5. $(-49)^{3/2}$

6. $(-27)^{1/3}$

7. $\left(\frac{16}{49}\right)^{-1/2}$

8. $x^{-3/5} \cdot x^{1/5}$

9. $a^{1/3} \cdot a^{4/3}$

10. $x \cdot x^{-1/4}$

11. $\frac{a^{3n}}{a^{-n}}$

12. $\left(a^{-1/4}\right)^{-4}$

13. $x^{2/5}x^{-3/5}$

14. $\left(a^4b^8\right)^{3/4}$

15. $\left(x^{-6}y^9\right)^{-1/3}$

16. $\left(\frac{x^{1/3}}{y^{-2}}\right)^6$

17. $x^{2/3}\left(x^{2/3} - x^{-1/3}\right)$

18. $\frac{a^{3/2}}{a^{1/2}}$

19. $\frac{x^{3n}}{x^n}$

20. $a^{n/2}a^{-n/4}$

21. $\frac{x^{-2/5}}{x^{1/5}}$

22. $a^{2n}a^{-4n}$

23. $\frac{x^{3/4}}{x^{-1/4}}$

24. $\left(x^{-3}\right)^{1/3}$

25. $\left(a^6\right)^{-2/3}$

26. $\left(x^{-2/3}\right)^9$

27. $\left(x^{-3/4}\right)^{12}$

1. ______
2. ______
3. ______
4. ______
5. ______
6. ______
7. ______
8. ______
9. ______
10. ______
11. ______
12. ______
13. ______
14. ______
15. ______
16. ______
17. ______
18. ______
19. ______
20. ______
21. ______
22. ______
23. ______
24. ______
25. ______
26. ______
27. ______

Name Score

**Rewrite the exponential expression as a radical expression.**

**1.** $2^{1/3}$ **2.** $7^{1/2}$ **3.** $b^{3/5}$

**4.** $(2x)^{7/4}$ **5.** $-9a^{2/3}$ **6.** $(ab^2)^{1/3}$

**7.** $(x^2y^2)^{3/5}$ **8.** $(a^2b^2)^{2/3}$ **9.** $(3a-2b)^{1/2}$

**10.** $(3x-4)^{1/3}$ **11.** $(2x-1)^{2/3}$ **12.** $x^{-3/4}$

1. ____________
2. ____________
3. ____________
4. ____________
5. ____________
6. ____________
7. ____________
8. ____________
9. ____________
10. ____________
11. ____________
12. ____________

**Rewrite the exponential expression as a radical expression.**

**13.** $\sqrt{15}$ **14.** $\sqrt[3]{y}$ **15.** $\sqrt[4]{5y^5}$

**16.** $-\sqrt{2x^3}$ **17.** $-\sqrt[3]{3x^5}$ **18.** $-\sqrt[4]{2x^7}$

**19.** $-\sqrt[3]{4x^4}$ **20.** $2x\sqrt[3]{y^2}$ **21.** $3x\sqrt[4]{y}$

**22.** $\sqrt{a^2-8}$ **23.** $\sqrt{2-x^2}$ **24.** $\sqrt{x^2+y^2}$

13. ____________
14. ____________
15. ____________
16. ____________
17. ____________
18. ____________
19. ____________
20. ____________
21. ____________
22. ____________
23. ____________
24. ____________

Name Score

**Simplify.**

1. $\sqrt{x^{12}}$
2. $\sqrt{x^{10}}$
3. $-\sqrt{x^{14}}$
4. $-\sqrt{a^4}$
5. $-\sqrt[3]{x^{12}y^6}$
6. $-\sqrt[3]{a^6b^{15}}$
7. $-\sqrt[3]{a^3b^9}$
8. $\sqrt[5]{x^{10}y^{25}}$
9. $\sqrt[4]{x^{12}y^{20}}$
10. $\sqrt[3]{-a^9b^{12}}$
11. $\sqrt{144x^{12}y^{16}}$
12. $\sqrt{9a^4b^{10}}$
13. $\sqrt{81x^6y^4}$
14. $\sqrt[3]{64x^6}$
15. $\sqrt[3]{125x^{12}y^3}$
16. $\sqrt[3]{-x^{12}y^{21}}$
17. $\sqrt[3]{-8a^{18}b^9}$
18. $-\sqrt[4]{a^8b^{12}}$
19. $\sqrt[5]{x^{20}y^{30}}$
20. $\sqrt[5]{x^{15}y^{25}}$
21. $\sqrt[4]{16a^{16}b^{20}}$
22. $\sqrt[4]{625a^8b^{12}}$
23. $\sqrt[5]{-a^5b^{15}}$
24. $\sqrt[5]{243x^{10}y^{15}}$

1. ______
2. ______
3. ______
4. ______
5. ______
6. ______
7. ______
8. ______
9. ______
10. ______
11. ______
12. ______
13. ______
14. ______
15. ______
16. ______
17. ______
18. ______
19. ______
20. ______
21. ______
22. ______
23. ______
24. ______

Name Score

**Simplify.**

1. $\sqrt{x^4y^6z^7}$
2. $\sqrt{98a^4b^6}$
3. $\sqrt{50xy^6z^9}$
4. $\sqrt{75x^3y^4z^6}$
5. $\sqrt{-16x^5}$
6. $\sqrt{-x^3y^6}$
7. $\sqrt{x^5y^7z^9}$
8. $\sqrt{x^6y^8z^4}$
9. $\sqrt{x^6z^{12}}$
10. $\sqrt{192xy^7z^{10}}$
11. $\sqrt{128x^3y^5z^6}$
12. $\sqrt[3]{8x^6y^9z^{10}}$
13. $\sqrt[3]{-64x^3y^9}$
14. $\sqrt[4]{-16x^4y^8}$
15. $\sqrt[4]{-81x^3y^6}$
16. $\sqrt[3]{a^5b^7}$
17. $\sqrt[3]{a^6b^8}$
18. $\sqrt[3]{-8x^6y^4}$
19. $\sqrt[3]{-64x^6y^9}$
20. $\sqrt[3]{a^5b^6c^7}$
21. $\sqrt[3]{a^{12}b^9c^8}$
22. $\sqrt[4]{a^{12}b^9c^8}$
23. $\sqrt[4]{16x^7y^5}$
24. $\sqrt[4]{64x^{12}y^8}$

1. ____________
2. ____________
3. ____________
4. ____________
5. ____________
6. ____________
7. ____________
8. ____________
9. ____________
10. ____________
11. ____________
12. ____________
13. ____________
14. ____________
15. ____________
16. ____________
17. ____________
18. ____________
19. ____________
20. ____________
21. ____________
22. ____________
23. ____________
24. ____________

Name Score

**Simplify.**

1. $3\sqrt{x}-7\sqrt{x}$

2. $4\sqrt{x}+10\sqrt{x}$

3. $\sqrt{20}-\sqrt{45}$

4. $\sqrt{48}-\sqrt{27}$

5. $\sqrt{98x}-\sqrt{72x}$

6. $\sqrt{75x}+\sqrt{108x}$

7. $\sqrt{54a}-\sqrt{24a}$

8. $\sqrt{75a}+\sqrt{48a}$

9. $3\sqrt{3x^3}+5x\sqrt{12x}$

10. $x\sqrt{48xy}-\sqrt{12x^3y}$

11. $3\sqrt[3]{2a^5}-5a\sqrt[3]{16a^2}$

12. $3a\sqrt{18a^3b^5}+5b\sqrt{2a^5b^3}$

13. $5b\sqrt{a^3b^7}-2a\sqrt{ab^9}$

14. $\sqrt[3]{8}-\sqrt[3]{27}$

15. $\sqrt[3]{24}+\sqrt[3]{192}$

16. $3\sqrt[3]{2a^4}-4a\sqrt[3]{54a}$

17. $4y\sqrt{12y}+3\sqrt{48y^3}$

18. $\sqrt{4a^3}-\sqrt{16a^3}+\sqrt{64a^3}$

19. $3b\sqrt[4]{81a^5b}-2a\sqrt[4]{256ab^5}$

20. $\sqrt[3]{54x^4y^3}+y\sqrt[3]{250x^4}-xy\sqrt[3]{16x}$

21. $3a\sqrt[4]{32a^5}-5a^2\sqrt[4]{128a}+\sqrt[4]{2a^9}$

22. $5y\sqrt[4]{3x^9y}-3xy\sqrt[4]{48x^5y}-2\sqrt[4]{243x^9y^5}$

1. ____
2. ____
3. ____
4. ____
5. ____
6. ____
7. ____
8. ____
9. ____
10. ____
11. ____
12. ____
13. ____
14. ____
15. ____
16. ____
17. ____
18. ____
19. ____
20. ____
21. ____
22. ____

Name Score

**Simplify.**

1. $\sqrt{2}\sqrt{8}$
2. $\sqrt{6}\sqrt{8}$
3. $\sqrt{7}\sqrt{28}$
4. $\sqrt[3]{8}\sqrt[3]{32}$
5. $\sqrt{x^3y^5}\sqrt{xy^3}$
6. $\sqrt[3]{9x^4y^2}\sqrt[3]{6x^2y^7}$
7. $\sqrt[3]{8a^3b^4}\sqrt[3]{16ab^5}$
8. $\sqrt{2}\left(\sqrt{50}-\sqrt{2}\right)$
9. $\sqrt{8}\left(\sqrt{2}-\sqrt{8}\right)$
10. $\sqrt{x}\left(\sqrt{x}+\sqrt{3}\right)$
11. $\sqrt{x}\left(\sqrt{x}-\sqrt{2}\right)$
12. $\sqrt{4x}\left(\sqrt{32x}-\sqrt{18}\right)$
13. $\left(\sqrt{2x}+3\right)^2$
14. $\left(\sqrt{x}+2\right)^2$
15. $\left(\sqrt{3x}-6\right)^2$
16. $3\sqrt{2x^2}\cdot 4\sqrt{8xy^2}\cdot\sqrt{3x^3y^4}$
17. $3\sqrt{15x^2y}\cdot 4\sqrt{5xy^2}\cdot 5\sqrt{3x^3y^5}$
18. $\sqrt[3]{9a^2b}\sqrt[3]{8ab}\sqrt[3]{3a^3b^4}$
19. $\sqrt[3]{3ab^2}\sqrt[3]{6a^2b}\sqrt[3]{9a^5b^6}$
20. $\left(\sqrt{2}-1\right)\left(\sqrt{2}+2\right)$
21. $\left(\sqrt{3}-3\right)\left(\sqrt{3}+1\right)$
22. $\left(\sqrt{x}-3\right)\left(\sqrt{x}+3\right)$
23. $\left(\sqrt{x}-4\right)\left(\sqrt{x}+4\right)$

1. ______________
2. ______________
3. ______________
4. ______________
5. ______________
6. ______________
7. ______________
8. ______________
9. ______________
10. ______________
11. ______________
12. ______________
13. ______________
14. ______________
15. ______________
16. ______________
17. ______________
18. ______________
19. ______________
20. ______________
21. ______________
22. ______________
23. ______________

Name Score

**Simplify.**

1. $\dfrac{\sqrt{48x^5}}{\sqrt{3x^3}}$

2. $\dfrac{\sqrt{24a^5b^3}}{\sqrt{12ab}}$

3. $\dfrac{\sqrt{27a^3b^5}}{\sqrt{3ab}}$

4. $\dfrac{1}{\sqrt{3}}$

5. $\dfrac{1}{\sqrt{5}}$

6. $\dfrac{1}{\sqrt{6x}}$

7. $\dfrac{3}{\sqrt{2y}}$

8. $\dfrac{9}{\sqrt{9x}}$

9. $\dfrac{4}{\sqrt{6a}}$

10. $\sqrt{\dfrac{5}{x}}$

11. $\sqrt{\dfrac{a}{2}}$

12. $\dfrac{2}{\sqrt[3]{3}}$

13. $\dfrac{3}{\sqrt[3]{9}}$

14. $\dfrac{\sqrt{20x^5y}}{\sqrt{40x^2y^2}}$

15. $\dfrac{\sqrt{18ab^3}}{\sqrt{48a^5b}}$

16. $\dfrac{\sqrt{24x^7y}}{\sqrt{40xy}}$

17. $\dfrac{3}{\sqrt{5}+1}$

18. $\dfrac{3}{3-\sqrt{5}}$

19. $\dfrac{2}{\sqrt{x}-4}$

20. $\dfrac{-5}{\sqrt{x}-5}$

21. $\dfrac{\sqrt{5}-\sqrt{2}}{\sqrt{5}+\sqrt{2}}$

22. $\dfrac{3\sqrt{32}-\sqrt{2}}{\sqrt{2}}$

23. $\dfrac{5\sqrt{125}+3\sqrt{80}}{\sqrt{5}}$

24. $\dfrac{\sqrt{x^3}+\sqrt{x}}{\sqrt{x}}$

25. $\dfrac{\sqrt{x^7}+\sqrt{x^5}}{\sqrt{x^5}}$

26. $\dfrac{9\sqrt{3x^3}-12x\sqrt{6x}}{\sqrt{6x}}$

27. $\dfrac{\sqrt{8x^3}-\sqrt{32x}}{\sqrt{2x}}$

1. __________
2. __________
3. __________
4. __________
5. __________
6. __________
7. __________
8. __________
9. __________
10. __________
11. __________
12. __________
13. __________
14. __________
15. __________
16. __________
17. __________
18. __________
19. __________
20. __________
21. __________
22. __________
23. __________
24. __________
25. __________
26. __________
27. __________

Name Score

**State the domain of the radical function in set builder notation.**

**1.** $f(x)=\sqrt{x-5}$

**2.** $g(x)=\frac{3}{4}\sqrt{4-x}+3$

**3.** $r(x)=5x\sqrt{x}+1$

**4.** $h(x)=-2x^{\frac{1}{2}}+3$

**5.** $C(x)=-2\sqrt[3]{x+3}$

**6.** $F(x)=4x^{\frac{3}{4}}-6$

1. ______________

2. ______________

3. ______________

4. ______________

5. ______________

6. ______________

**State the domain of the radical function in interval notation.**

**7.** $G(x)=\sqrt[3]{7x}$

**8.** $V(x)=2x^{\frac{1}{2}}-5$

**9.** $h(x)=3x\sqrt{1-x}$

**10.** $Y(x)=-2\sqrt{3x+6}$

**11.** $R(x)=\frac{1}{4}\sqrt[3]{5-x}$

**12.** $C(x)=\sqrt[4]{(3+x)^3}$

7. ______________

8. ______________

9. ______________

10. ______________

11. ______________

12. ______________

Name Score

**Sketch a graph of the radical function.**

1. $f(x) = \sqrt{x+3}$

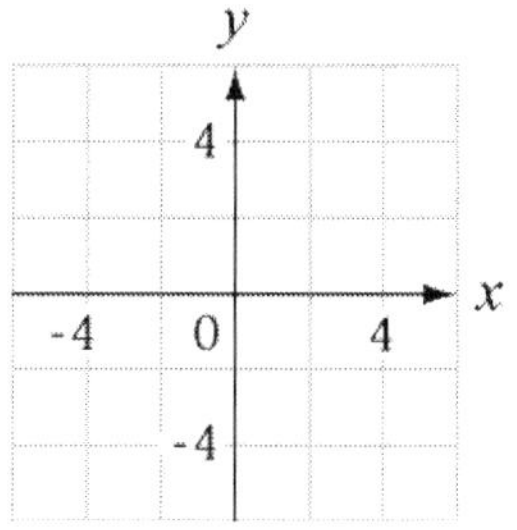


2. $Q(x) = -\sqrt[3]{x-1}$

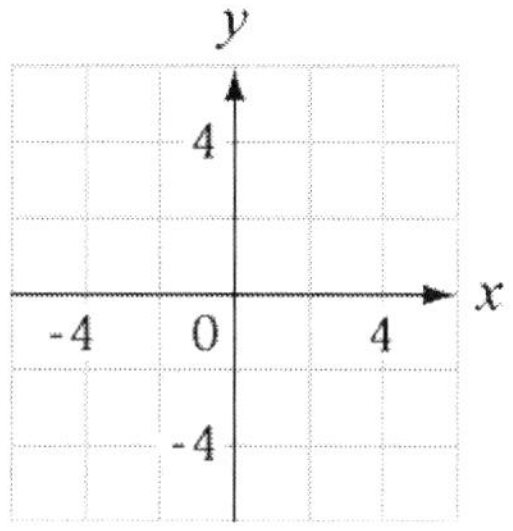


3. $r(x) = \sqrt[3]{x-2}$

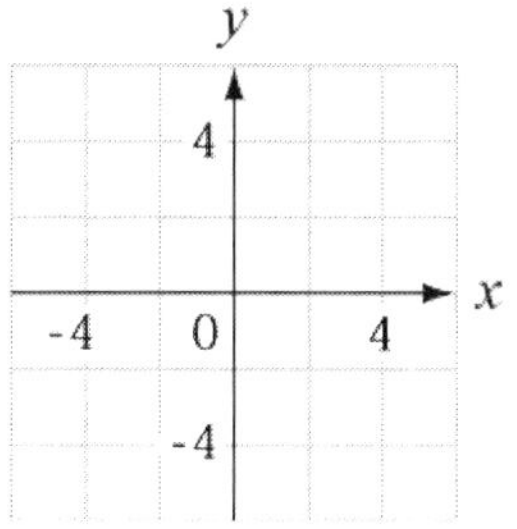


4. $g(x) = -\sqrt{x-2}$

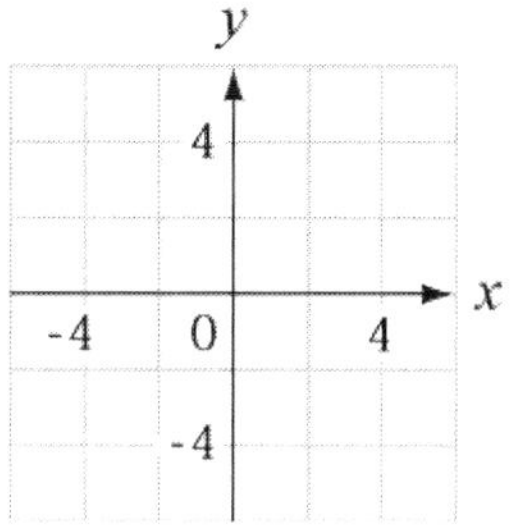


5. $h(x) = -\sqrt{x+1}$

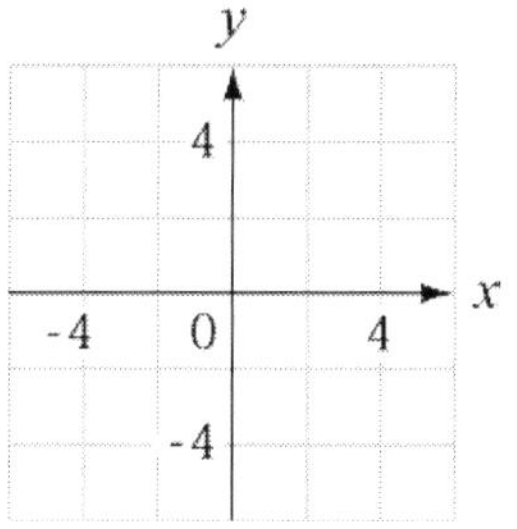


6. $M(x) = -\sqrt[3]{x+1}$

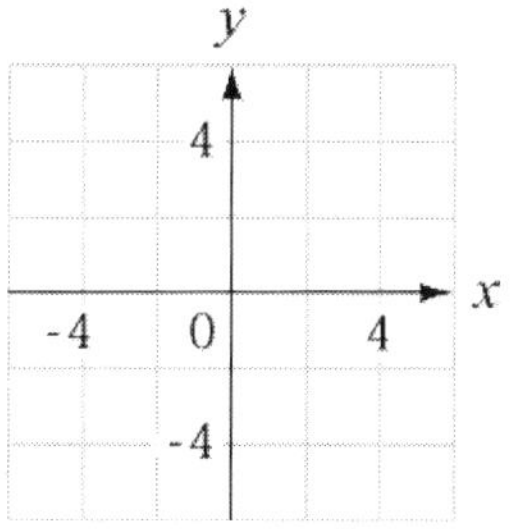


7. $A(x) = \sqrt[3]{x+2}$

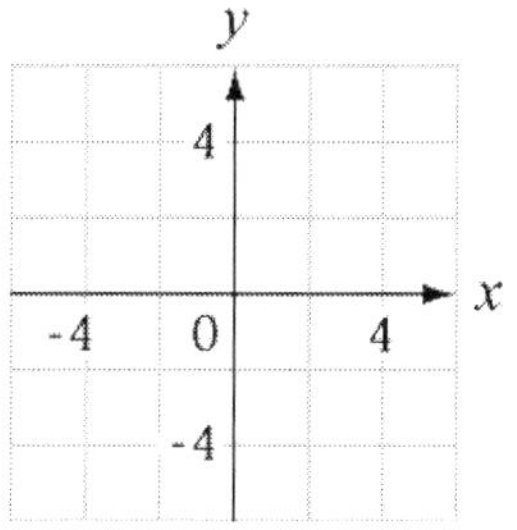


8. $C(x) = \sqrt{x-3}$

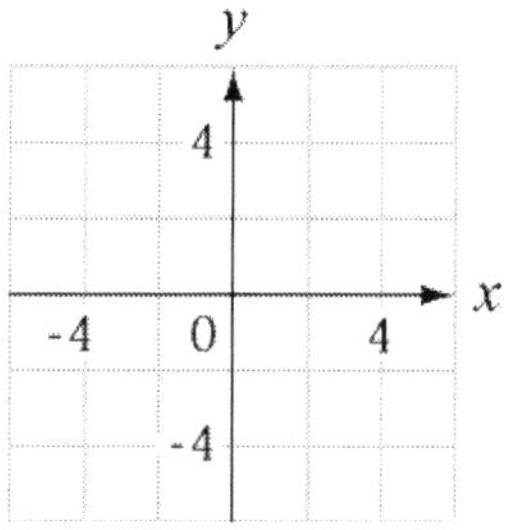

Name Score

**Solve.**

1. $\sqrt{x} = 3$

2. $\sqrt{a} = 7$

3. $\sqrt{x} = 6$

4. $\sqrt[3]{a} = 2$

5. $\sqrt[3]{x} = 6$

6. $\sqrt{3y} = 12$

7. $\sqrt[3]{3x} = -6$

8. $\sqrt{3x} = -6$

9. $\sqrt{3x+1} - 7 = 0$

10. $\sqrt{5x+1} - 4 = 0$

11. $\sqrt{3x+5} = \sqrt{4x-1}$

12. $\sqrt{2x+3} = \sqrt{5x-9}$

13. $\sqrt[3]{3x+4} = 4$

14. $\sqrt[3]{2x-6} = \sqrt[3]{x+2}$

15. $\sqrt[3]{x-10} = \sqrt[3]{5x+18}$

16. $\sqrt[4]{3x-5} = 2$

17. $\sqrt{3x-2} + 3 = 5$

18. $\sqrt[3]{2x-1} + 4 = 3$

1. ________
2. ________
3. ________
4. ________
5. ________
6. ________
7. ________
8. ________
9. ________
10. ________
11. ________
12. ________
13. ________
14. ________
15. ________
16. ________
17. ________
18. ________

Name Score

**Solve.**

1. Find the length of a rectangle that has a diagonal of 13 ft and a width of 5 ft.

2. A 20-foot ladder is leaning against a building. How high on the building will the ladder reach when the bottom of the ladder is 7 ft from the building? Round to the nearest tenth.

3. How far would a submarine periscope have to be above the water to locate a ship 4.9 mi away? The equation for the distance in miles that the lookout can see is $d = 1.4\sqrt{h}$, where $h$ is the height in feet above the surface of the water.

4. How far would a submarine periscope have to be above the water to locate a ship 5.4 mi away? The equation for the distance in miles that the lookout can see is $d = 1.4\sqrt{h}$, where $h$ is the height in feet above the surface of the water. Round to the nearest hundredth.

5. An object is dropped from a tall tower. Find the distance the object has fallen when the speed reaches 75 ft/s. Use the equation $v = \sqrt{64d}$, where $v$ is the speed of the object and $d$ is the distance. Round to the nearest hundredth.

6. An object is dropped from a tall tower. Find the distance the object has fallen when the speed reaches 320 ft/s. Use the equation $v = \sqrt{64d}$, where $v$ is the speed of the object and $d$ is the distance.

7. Find the distance required for a car to reach a velocity of 90 ft/s when the acceleration is 25 ft/s$^2$. Use the equation $v = \sqrt{2as}$, where $v$ is the velocity, $a$ is the acceleration, and $s$ is the distance.

8. Find the distance required for a car to reach a velocity of 80 ft/s when the acceleration is 16 ft/s$^2$. Use the equation $v = \sqrt{2as}$, where $v$ is the velocity, $a$ is the acceleration, and $s$ is the distance.

9. Find the length of a pendulum that makes one swing in 2 s. The equation for the time of one swing of a pendulum is given by $T = 2\pi\sqrt{\frac{L}{32}}$, where $T$ is the time in seconds and $L$ is the length in feet. Use 3.14 for $\pi$. Round to the nearest hundredth.

10. Find the length of a pendulum that makes one swing in 3.2 s. The equation for the time of one swing of a pendulum is given by $T = 2\pi\sqrt{\frac{L}{32}}$, where $T$ is the time in seconds and $L$ is the length in feet. Use 3.14 for $\pi$. Round to the nearest hundredth.

1. ______
2. ______
3. ______
4. ______
5. ______
6. ______
7. ______
8. ______
9. ______
10. ______

Name Score

**Simplify.**

1. $\sqrt{-49}$

2. $\sqrt{-8}$

3. $\sqrt{-12x^2}$

4. $\sqrt{-200}$

5. $\sqrt{-128}$

6. $\sqrt{-144}$

7. $\sqrt{-108x}$

8. $\sqrt{-180y^3}$

9. $\sqrt{9a^6}+\sqrt{-9a^6}$

10. $\sqrt{16}+\sqrt{-25}$

11. $\sqrt{36}+\sqrt{-16}$

12. $\sqrt{49x^2y^4}+\sqrt{-100x^4y^2}$

13. $\sqrt{48}-\sqrt{-27}$

14. $\sqrt{50xy^2}+\sqrt{-18x^2y}$

15. $\sqrt{24x^5y^4z^3}-\sqrt{-36x^2y^8z^6}$

16. $\sqrt{50}-\sqrt{-98}$

17. $\sqrt{150}-\sqrt{-27}$

18. $\sqrt{72x^5z^2}-\sqrt{-128y^6}$

1. ______
2. ______
3. ______
4. ______
5. ______
6. ______
7. ______
8. ______
9. ______
10. ______
11. ______
12. ______
13. ______
14. ______
15. ______
16. ______
17. ______
18. ______

Name Score

**Add or subtract.**

**1.** $(5-8i)+(6+2i)$

**2.** $(4-6i)+(7-3i)$

**3.** $(4+6i)+(8-3i)$

**4.** $(7-4i)+(5+3i)$

**5.** $(4-6i)+(9-i)$

**6.** $\left(4-\sqrt{-16}\right)-\left(10-\sqrt{-4}\right)$

**7.** $\left(6-\sqrt{-100}\right)-\left(12-\sqrt{-121}\right)$

**8.** $\left(8-\sqrt{-27}\right)-\left(7+\sqrt{-48}\right)$

**9.** $\left(\sqrt{32}+\sqrt{-8}\right)+\left(\sqrt{50}-\sqrt{-72}\right)$

**10.** $\left(\sqrt{24}-\sqrt{-20}\right)-\left(\sqrt{96}+\sqrt{-45}\right)$

**11.** $(9-2i)+i$

**12.** $(8-5i)+2i$

**13.** $(6-4i)+3i$

**14.** $(7-6i)+5i$

**15.** $(5+4i)+(-5-4i)$

**16.** $(7-5i)+(-10+2i)$

**17.** $(8+3i)+(-6-4i)$

**18.** $(9-2i)+(-4+4i)$

1. ____________
2. ____________
3. ____________
4. ____________
5. ____________
6. ____________
7. ____________
8. ____________
9. ____________
10. ____________
11. ____________
12. ____________
13. ____________
14. ____________
15. ____________
16. ____________
17. ____________
18. ____________

Name Score

**Multiply.**

1. $(5i)(-8i)$
2. $(-5i)(-3i)$
3. $(-8i)(20)$
4. $(-4i)(-8i)$
5. $\sqrt{-3}\sqrt{-27}$
6. $\sqrt{-2}\sqrt{-32}$
7. $\sqrt{-6}\sqrt{-8}$
8. $3i(5+i)$
9. $4i(2+3i)$
10. $(5-2i)(3+i)$
11. $(4+3i)(3+i)$
12. $(5-i)^2$
13. $(5-i)(2-i)$
14. $(3-4i)(2-i)$
15. $(6+2i)(6-2i)$
16. $(3-5i)(3+5i)$
17. $(5-3i)(2+i)$
18. $(3-6i)(3-i)$
19. $(5+4i)(2+i)$
20. $(5+2i)^2$
21. $(2-i)(2+i)$
22. $\left(\frac{1}{2}+2i\right)\left(2+\frac{1}{2}i\right)$
23. $(3-6i)(2+5i)$
24. $(3-i)\left(\frac{1}{3}+\frac{2}{3}i\right)$

1. ____
2. ____
3. ____
4. ____
5. ____
6. ____
7. ____
8. ____
9. ____
10. ____
11. ____
12. ____
13. ____
14. ____
15. ____
16. ____
17. ____
18. ____
19. ____
20. ____
21. ____
22. ____
23. ____
24. ____

Name Score

**Divide.**

1. $\frac{4}{i}$

2. $\frac{2}{3i}$

3. $\frac{3}{4i}$

4. $\frac{4-6i}{-2i}$

5. $\frac{15+2i}{-5i}$

6. $\frac{5}{4+i}$

7. $\frac{8}{3+i}$

8. $\frac{5}{5+2i}$

9. $\frac{4}{4-i}$

10. $\frac{6}{5-i}$

11. $\frac{1-5i}{5+i}$

12. $\frac{1-4i}{4+i}$

13. $\frac{3+6i}{4+i}$

14. $\frac{4+20i}{5+i}$

15. $\frac{6-3i}{3-i}$

16. $\frac{4+8i}{2-i}$

17. $\frac{\sqrt{-12}}{\sqrt{3}-\sqrt{-6}}$

18. $\frac{\sqrt{-2}}{\sqrt{8}-\sqrt{-2}}$

19. $\frac{\sqrt{-3}}{\sqrt{12}-\sqrt{-6}}$

20. $\frac{4-5i}{5+i}$

21. $\frac{3-2i}{2+i}$

22. $\frac{4+3i}{2-i}$

23. $\frac{2+5i}{1-i}$

24. $\frac{6+5i}{2-i}$

1. ________
2. ________
3. ________
4. ________
5. ________
6. ________
7. ________
8. ________
9. ________
10. ________
11. ________
12. ________
13. ________
14. ________
15. ________
16. ________
17. ________
18. ________
19. ________
20. ________
21. ________
22. ________
23. ________
24. ________

Name Score

**Solve by factoring.**

1. $y^2+5y=0$
2. $t^2-16=0$
3. $p^2-64=0$
4. $s^2+s-12=0$
5. $y^2-4y+4=0$
6. $x^2+8x+16=0$
7. $8z^2-24z=0$
8. $3y^2+15y=0$
9. $t^2-2t=15$
10. $p^2+3p=4$
11. $v^2+12=8v$
12. $t^2-12=11t$
13. $3x^2+x-2=0$
14. $2x^2-5x-3=0$
15. $3x^2-7x-6=0$
16. $t+18=t(t+8)$
17. $x+12=x(x-3)$
18. $t+3=t(t+3)$
19. $(2v-3)(3v+1)=5v^2-9v+12$
20. $u^2-3u+9=(u+1)(2u+3)$

1. ____________
2. ____________
3. ____________
4. ____________
5. ____________
6. ____________
7. ____________
8. ____________
9. ____________
10. ____________
11. ____________
12. ____________
13. ____________
14. ____________
15. ____________
16. ____________
17. ____________
18. ____________
19. ____________
20. ____________

**Solve for *x* by factoring.**

21. $x^2+4bx-21b^2=0$
22. $x^2-5bx+4b^2=0$
23. $x^2-7cx-8c^2=0$
24. $x^2-ax-12a^2=0$
25. $2x^2-5cx+2c^2=0$
26. $3x^2+4ax-6a^2=0$

21. ____________
22. ____________
23. ____________
24. ____________
25. ____________
26. ____________

Name Score

**Solve by taking square roots.**

**1.** $y^2 = 25$

**2.** $x^2 = 36$

**3.** $z^2 = -9$

**4.** $v^2 = -49$

**5.** $s^2 - 1 = 0$

**6.** $t^2 - 9 = 0$

**7.** $z^2 + 25 = 0$

**8.** $u^2 - 27 = 0$

**9.** $t^2 + 50 = 0$

**10.** $2(x+3)^2 = 32$

**11.** $(x-2)^2 = 16$

**12.** $6(y+1)^2 = 24$

**13.** $2(x-2)^2 = 162$

**14.** $4(z-1)^2 = 100$

**15.** $\left(v - \frac{1}{3}\right)^2 = \frac{1}{9}$

**16.** $\left(r + \frac{1}{2}\right)^2 = \frac{1}{4}$

**17.** $(v-5)^2 + 15 = 0$

**18.** $(x+4)^2 + 28 = 0$

**19.** $(x-1)^2 - 18 = 0$

**20.** $(x+2)^2 + 16 = 0$

**21.** $(x-3)^2 - 50 = 0$

**22.** $\left(x - \frac{3}{5}\right)^2 - 75 = 0$

**23.** $\left(u - \frac{1}{3}\right)^2 - 27 = 0$

**24.** $\left(x + \frac{1}{2}\right)^2 - 45 = 0$

**1.** ____________
**2.** ____________
**3.** ____________
**4.** ____________
**5.** ____________
**6.** ____________
**7.** ____________
**8.** ____________
**9.** ____________
**10.** ____________
**11.** ____________
**12.** ____________
**13.** ____________
**14.** ____________
**15.** ____________
**16.** ____________
**17.** ____________
**18.** ____________
**19.** ____________
**20.** ____________
**21.** ____________
**22.** ____________
**23.** ____________
**24.** ____________

Name Score

**Solve by completing the square.**

**1.** $v^2 + 6v - 7 = 0$

**2.** $w^2 - 2w - 16 = 0$

**3.** $z^2 - 4z + 4 = 0$

**4.** $u^2 + 8u + 16 = 0$

**5.** $x^2 - 6x + 5 = 0$

**6.** $y^2 + 8y + 10 = 0$

**7.** $z^2 - 2z - 3 = 0$

**8.** $x^2 + 3x - 10 = 0$

**9.** $v^2 - 5v - 36 = 0$

**10.** $y^2 - 7y + 12 = 0$

**11.** $y^2 - 9y + 18 = 0$

**12.** $x^2 + 4x - 16 = 0$

**13.** $x^2 - 6x + 2 = 0$

**14.** $3x^2 + 2x - 1 = 0$

**15.** $2y^2 - 3y + 1 = 0$

**16.** $2x^2 - 5x - 12 = 0$

**17.** $x^2 = 10x + 24$

**18.** $x^2 = 6x + 16$

**19.** $z^2 = 4z + 5$

**20.** $10x^2 - 9x + 2 = 0$

**21.** $4x^2 - 8x + 3 = 0$

**22.** $3x^2 + 14x + 12 = 0$

**23.** $6x^2 + 7x - 3 = 0$

**24.** $4x^2 - 12x - 7 = 0$

**25.** $2x^2 = 10 - x$

**26.** $(x+3)(x-1) = x - 2$

**27.** $(y-2)^2 = 3y - 6$

**1.** ______
**2.** ______
**3.** ______
**4.** ______
**5.** ______
**6.** ______
**7.** ______
**8.** ______
**9.** ______
**10.** ______
**11.** ______
**12.** ______
**13.** ______
**14.** ______
**15.** ______
**16.** ______
**17.** ______
**18.** ______
**19.** ______
**20.** ______
**21.** ______
**22.** ______
**23.** ______
**24.** ______
**25.** ______
**26.** ______
**27.** ______

Name Score

**Solve by using the quadratic formula.**

1. $w^2 = 5w + 50$
2. $t^2 = 3t + 40$
3. $x^2 = 9 - 8x$
4. $2x^2 - 5x + 6 = 0$
5. $8x^2 + 10x = 3$
6. $6x^2 + 7x - 3 = 0$
7. $2x^2 - 3x + 1 = 0$
8. $4x^2 + 20x + 9 = 0$
9. $2x^2 + 2x + 5 = 0$
10. $t^2 - 4t + 6 = 0$
11. $3x^2 - 6x - 1 = 0$
12. $12x^2 - 5x - 3 = 0$
13. $2x^2 - 11x + 5 = 0$
14. $6x^2 - 11x + 9 = 0$
15. $3x^2 - 3x + 4 = 0$
16. $5y(y+3) = 4y + 12$
17. $(2x-3)(x-1) = 3$
18. $(3t+1)(t-2) = 10$

1. ____________
2. ____________
3. ____________
4. ____________
5. ____________
6. ____________
7. ____________
8. ____________
9. ____________
10. ____________
11. ____________
12. ____________
13. ____________
14. ____________
15. ____________
16. ____________
17. ____________
18. ____________

**Solve by using the quadratic equation. Approximate solutions to the nearest thousandths.**

19. $p^2 - 6p + 2 = 0$
20. $w^2 - 8w + 3 = 0$
21. $r^2 - 7r + 4 = 0$

19. ____________
20. ____________
21. ____________

**Use the discriminant to determine whether the quadratic equation has one real number solution, two real number solutions, or two complex number solutions.**

22. $9x^2 + 30x + 25 = 0$
23. $2p^2 + 7p + 3 = 0$
24. $3z^2 + 1 = 0$

22. ____________
23. ____________
24. ____________

Name Score

**Solve.**

1. $z^4 - 5z^2 + 4 = 0$

2. $x^4 - 20x^2 + 64 = 0$

3. $y^4 - 17y^2 + 16 = 0$

4. $x^4 - 29x^2 + 100 = 0$

5. $x^4 - 8x^2 - 9 = 0$

6. $x^4 - 20x^2 + 64 = 0$

7. $x - x^{1/2} - 20 = 0$

8. $w - 20w^{1/2} - 8 = 0$

9. $w - 5w^{1/2} + 4 = 0$

10. $x^4 + 4x^2 - 32 = 0$

11. $x^4 - 7x^2 + 12 = 0$

12. $x^4 + x^2 - 12 = 0$

13. $x^4 - 6x^2 + 8 = 0$

14. $x^4 - 1 = 0$

15. $x^4 - 16 = 0$

16. $x^4 - 64 = 0$

17. $x^4 - 16x^2 + 63 = 0$

18. $x - 5x^{1/2} + 6 = 0$

19. $x - 9x^{1/2} + 14 = 0$

20. $x - 9x^{1/2} + 8 = 0$

21. $y^{2/3} - 5y^{1/3} + 4 = 0$

22. $4y^4 - 17y^2 + 4 = 0$

23. $16w^4 - 8w^2 + 1 = 0$

24. $z^{2/3} - 1 = 0$

1. ____________
2. ____________
3. ____________
4. ____________
5. ____________
6. ____________
7. ____________
8. ____________
9. ____________
10. ____________
11. ____________
12. ____________
13. ____________
14. ____________
15. ____________
16. ____________
17. ____________
18. ____________
19. ____________
20. ____________
21. ____________
22. ____________
23. ____________
24. ____________

Name Score

**Solve.**

**1.** $x = \sqrt{x} + 2$

**2.** $\sqrt{3x+1} = x - 1$

**3.** $\sqrt{2s-1} = s - 2$

**4.** $\sqrt{3x+1} = x - 9$

**5.** $\sqrt{5x-4} = x$

**6.** $\sqrt{5x+4} = x - 2$

**7.** $\sqrt{2y-6} = 3 - y$

**8.** $\sqrt{x-2} = 4 - x$

**9.** $x - 1 = \sqrt{x+5}$

**10.** $\sqrt{x-7} = \sqrt{x} - 1$

**11.** $x - 3 = \sqrt{x-3}$

**12.** $x - 2 = \sqrt{x+4}$

**13.** $\sqrt{3x+1} = x - 3$

**14.** $\sqrt{2x-5} + 2 = x$

**15.** $\sqrt{x+2} = x - 4$

**16.** $\sqrt{x+7} + 1 = \sqrt{x+12}$

**17.** $\sqrt{4-x} + \sqrt{x+6} = 4$

**18.** $\sqrt{x+5} + \sqrt{x} = 5$

**19.** $\sqrt{y} + 3 = \sqrt{y+27}$

**20.** $\sqrt{x+2} - \sqrt{x-1} = 1$

**21.** $\sqrt{2x+4} = \sqrt{x} - 2$

**22.** $\sqrt{3x-2} - \sqrt{2x-3} = 1$

**23.** $\sqrt{3-2x} + \sqrt{2x+7} = 4$

**1.** ______________
**2.** ______________
**3.** ______________
**4.** ______________
**5.** ______________
**6.** ______________
**7.** ______________
**8.** ______________
**9.** ______________
**10.** ______________
**11.** ______________
**12.** ______________
**13.** ______________
**14.** ______________
**15.** ______________
**16.** ______________
**17.** ______________
**18.** ______________
**19.** ______________
**20.** ______________
**21.** ______________
**22.** ______________
**23.** ______________

Name Score

**Solve.**

1. $z = \frac{6}{z-5}$

2. $\frac{z}{3z-4} = \frac{3}{z+2}$

3. $\frac{3x}{x+2} = \frac{6}{x-1}$

4. $\frac{4x}{x-2} = \frac{12}{x-4}$

5. $\frac{x-3}{x+4} + x = 3$

6. $\frac{4v}{v-3} = \frac{7}{v+8}$

7. $\frac{x+3}{x-1} = \frac{x-1}{x+1}$

8. $\frac{12}{x-1} + \frac{12}{x+1} = 5$

9. $\frac{3x-2}{x-3} + x = 14$

10. $\frac{4x-1}{x-4} - x = 2$

11. $\frac{3v+4}{v+5} - 4v = 8$

12. $\frac{2x+1}{x-1} - x = -1$

13. $\frac{8}{3x+2} + \frac{2}{x} = -3$

14. $\frac{14}{z-3} + \frac{3}{z-3} = -5$

15. $\frac{5x-3}{x+6} - \frac{2x+1}{x+4} = -1$

16. $\frac{3x-1}{x+5} + \frac{2x-3}{x+2} = 4$

17. $\frac{2w}{w+3} + \frac{1}{w+1} = 2$

18. $\frac{3v}{v+3} + \frac{4}{v+5} = 2$

19. $\frac{x+4}{x+2} - \frac{x-1}{x+3} = 4$

1. ______
2. ______
3. ______
4. ______
5. ______
6. ______
7. ______
8. ______
9. ______
10. ______
11. ______
12. ______
13. ______
14. ______
15. ______
16. ______
17. ______
18. ______
19. ______

Name Score

**Solve.**

**1.** The length of a rectangle is 2 ft less than three times the width. The area of the rectangle is 40 $ft^2$. Find the length and width of the rectangle.

**2.** The height of a triangle is 3 m less than the length of the base. The area of the triangle is 20 $m^2$. Find the height of the triangle and the length of its base.

**3.** The difference between the squares of two consecutive integers is eleven. Find the two integers.

**4.** The sum of the squares of two consecutive odd integers is thirty-four. Find the two integers.

**5.** Five times an integer plus three times the square of the integer is twenty-two. Find the integer.

**6.** The difference between the squares of two consecutive odd integers is thirty-two. Find the two integers.

**7.** An old pump requires 9 h longer to empty a pool than does a new pump. With both pumps working, the pool can be emptied in 6 h. Find the time required for the new pump, working alone, to empty the pool.

**8.** A small air conditioner requires 16 min longer to cool a room 4° than does a larger air conditioner. Working together, the two air conditioners can cool the room 4° in 6 min. How long would it take each air conditioner, working alone, to cool the room 4°?

**9.** An old copier takes 15 min longer to copy a report than does a second, new model. With both copiers working, the report can be printed in 18 min. How long would it take each copier, working alone, to copy the report?

**10.** A car travels 240 mi. A second car, traveling 12 mph faster than the first, makes the same trip in 1 h less time. Find the speed of each car.

**1.** ______________

**2.** ______________

**3.** ______________

**4.** ______________

**5.** ______________

**6.** ______________

**7.** ______________

**8.** ______________

**9.** ______________

**10.** ______________

Name Score

**Find the vertex and axis of symmetry of the parabola. Then sketch the graph.**

**1.** $y = x^2 - 1$

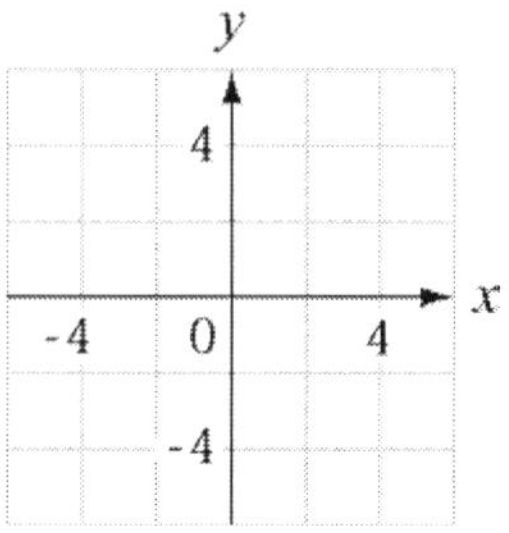


**2.** $y = \frac{1}{2}x^2 + 2$

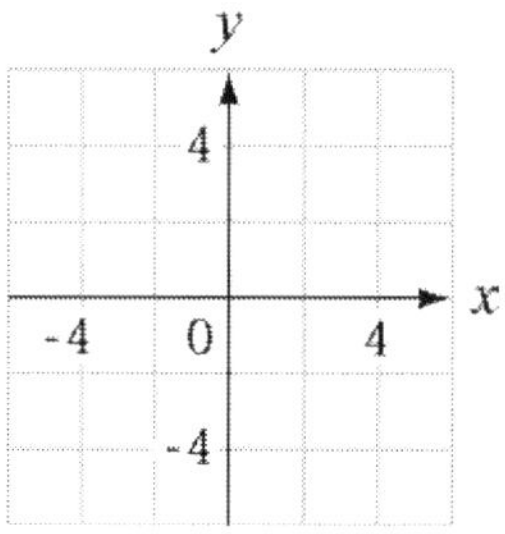


**3.** $y = x^2 + 3x$

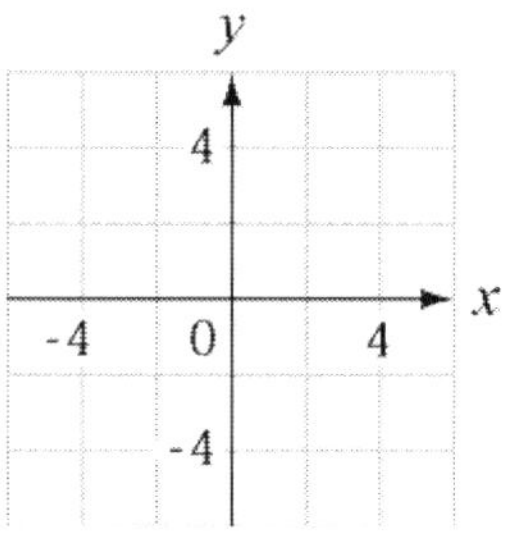


**4.** $y = -x^2 + 3x - 2$

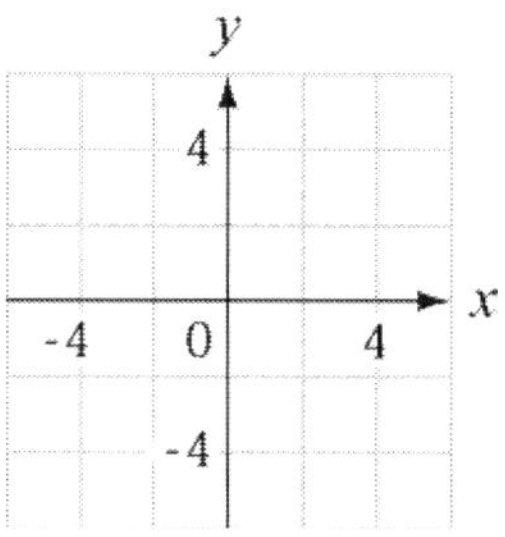


**1.** __________

**2.** __________

**3.** __________

**4.** __________

**Graph the function. State the domain and range of the function.**

**5.** $y = x^2 + x - 2$

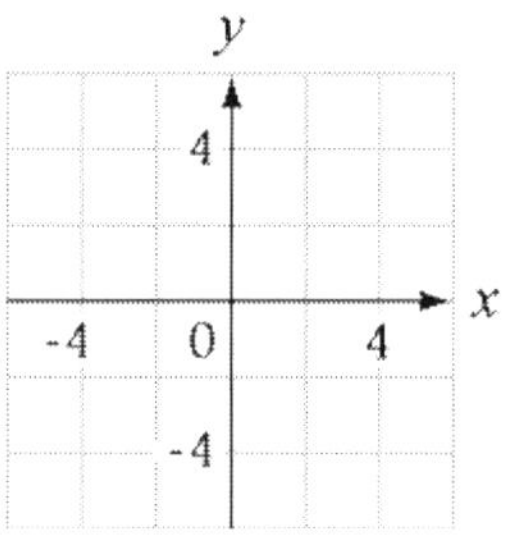


**6.** $y = -x^2 + 1$

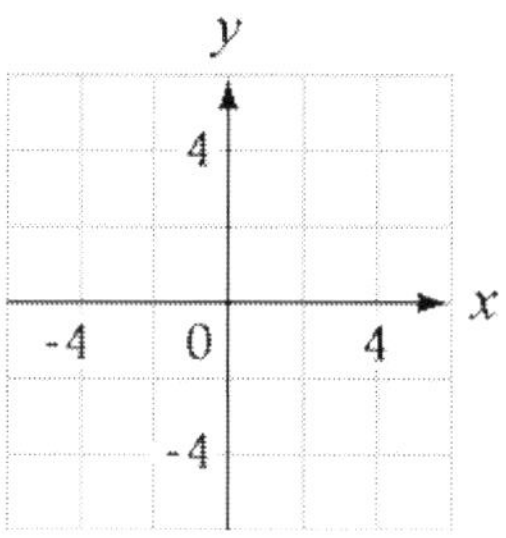


**7.** $y = 2x^2 + x - 3$

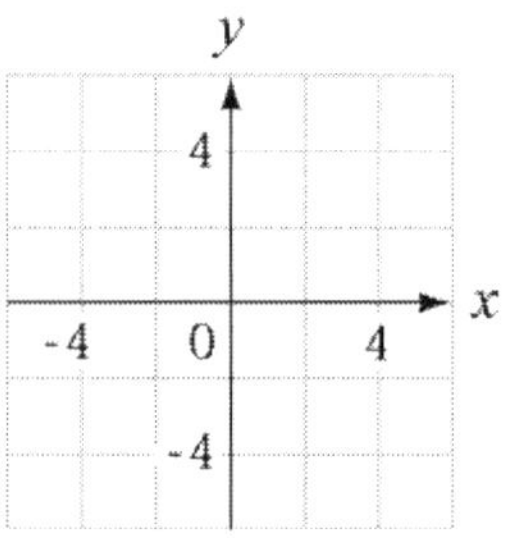


**8.** $y = x^2 - 3x$

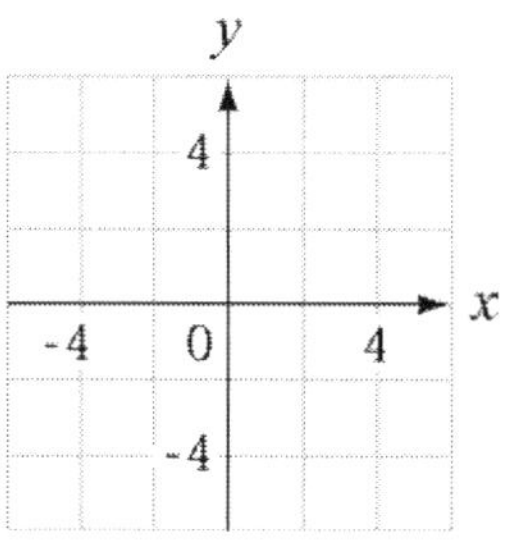


**5.** __________

**6.** __________

**7.** __________

**8.** __________

Name Score

**Find the $x$-intercepts of the parabola.**

1. $y = x^2 - 1$
2. $y = x^2 - 16$
3. $y = x^2 - x$
4. $y = 4x^2 - 8x$
5. $y = 2x^2 - 3x - 2$
6. $y = x^2 + 5x - 6$

1. ________
2. ________
3. ________
4. ________
5. ________
6. ________

**Find the zeros of the function.**

7. $f(x) = 3x^2 - 4x + 1$
8. $f(x) = x^2 - x - 12$
9. $f(x) = x^2 + 4x + 8$
10. $f(x) = -x^2 - 3x - 4$
11. $f(x) = x^2 - 3x - 4$
12. $f(x) = x^2 + 2x + 15$

7. ________
8. ________
9. ________
10. ________
11. ________
12. ________

**Use the discriminant to determine the number of $x$-intercepts of the parabola.**

13. $y = 3x^2 - 5x - 2$
14. $y = x^2 + x + 2$
15. $y = 2x^2 - 9x - 5$
16. $y = 3x^2 + 2x + 2$
17. $y = 2x^2 + x - 1$
18. $y = 2x^2 + 5x - 3$
19. $y = 3x^2 + x + 2$
20. $y = x^2 - 4x + 4$
21. $y = x^2 - 6x + 9$
22. $y = 4x^2 + x - 3$
23. $y = 2x^2 - 3x - 2$
24. $y = -2x^2 + 3x - 3$

13. ________
14. ________
15. ________
16. ________
17. ________
18. ________
19. ________
20. ________
21. ________
22. ________
23. ________
24. ________

Name Score

**Find the minimum or maximum value of the quadratic function.**

1. $f(x) = x^2 + 6x + 8$

2. $f(x) = x^2 + 2x - 3$

3. $f(x) = -2x^2 + 8x + 5$

4. $f(x) = -x^2 - 2x + 1$

5. $f(x) = x^2 + 6x - 2$

6. $f(x) = x^2 - 3x + 2$

7. $f(x) = -3x^2 - 12x + 1$

8. $f(x) = 2x^2 + 4x - 5$

9. $f(x) = -x^2 - x + 6$

10. $f(x) = x^2 - 4x - 21$

11. $f(x) = 2x^2 + 6x - 1$

12. $f(x) = -2x^2 + 4x - 1$

13. $f(x) = x^2 + 6x + 6$

14. $f(x) = x^2 - 3x + 6$

15. $f(x) = -3x^2 + 4x + 6$

16. $f(x) = -2x^2 - 3x + 1$

1. ________
2. ________
3. ________
4. ________
5. ________
6. ________
7. ________
8. ________
9. ________
10. ________
11. ________
12. ________
13. ________
14. ________
15. ________
16. ________

Name Score

**Solve.**

**1.** The height in feet ($s$) of a rock thrown straight up is given by the function $s(t) = -16t^2 + 96t$, where $t$ is the time in seconds. Find the maximum height above the ground that the rock will attain.

**2.** The height in feet ($s$) of a ball thrown upward at an initial speed of 72 ft/s from a platform 50 ft high is given by the function $s(t) = -16t^2 + 72t + 50$, where $t$ is the time in seconds. Find the maximum height above the ground that the ball will attain.

**3.** A pool is treated with a chemical to reduce the amount of algae. The amount of algae in the pool $t$ days after the treatment can be approximated by the function $A(t) = 30t^2 - 300t + 400$. How many days after treatment will the pool have the least amount of algae?

**4.** A manufacturer of video cassette recorders believes that the revenue the company receives is related to the price ($P$) of a recorder by the function $R(P) = 75P - \frac{1}{5}P^2$. What price will give the maximum revenue?

**5.** The height in feet ($s$) of a rocket after $t$ seconds is given by the formula $s(t) = 224t - 16t^2$. After how many seconds does the rocket reach its maximum height?

**6.** Find the two numbers whose sum is 50 and whose product is a maximum.

**7.** Find two numbers whose sum is 30 and whose product is a maximum.

**8.** Find two numbers whose difference is 20 and whose product is a minimum.

**9.** A rectangle has a perimeter of 56 ft. Find the dimensions of the rectangle with this perimeter that will have the maximum area. What is the maximum area?

**10.** The perimeter of a rectangular window is 32 ft. Find the dimensions of the window with this perimeter that will enclose the largest area. What is the maximum area?

**1.** ______

**2.** ______

**3.** ______

**4.** ______

**5.** ______

**6.** ______

**7.** ______

**8.** ______

**9.** ______

**10.** ______

Name Score

**Solve and graph the solution set.**

**1.** $(x-3)(x+3) > 0$

-5 -4 -3 -2 -1 0 1 2 3 4 5

**2.** $(x+4)(x+1) \ge 0$

-5 -4 -3 -2 -1 0 1 2 3 4 5

**3.** $x^2 - x - 6 < 0$

-5 -4 -3 -2 -1 0 1 2 3 4 5

**4.** $x^2 + 4x - 5 \le 0$

-5 -4 -3 -2 -1 0 1 2 3 4 5

**5.** $(x-4)(x+4)(x+3) < 0$

-5 -4 -3 -2 -1 0 1 2 3 4 5

**6.** $(x+5)(x+1)(x-3) > 0$

-5 -4 -3 -2 -1 0 1 2 3 4 5

**7.** $\dfrac{x+5}{x-1} > 0$

-5 -4 -3 -2 -1 0 1 2 3 4 5

**8.** $(x+3)(x-3)(x-2) \ge 0$

-5 -4 -3 -2 -1 0 1 2 3 4 5

**9.** $(x-2)(x+4)(x-1) \le 0$

-5 -4 -3 -2 -1 0 1 2 3 4 5

**10.** $\dfrac{x+3}{x} > 0$

-5 -4 -3 -2 -1 0 1 2 3 4 5

**11.** $x^2 \le 16$

-5 -4 -3 -2 -1 0 1 2 3 4 5

**12.** $\dfrac{2x}{x-3} > 1$

-5 -4 -3 -2 -1 0 1 2 3 4 5

**1.** ____________

**2.** ____________

**3.** ____________

**4.** ____________

**5.** ____________

**6.** ____________

**7.** ____________

**8.** ____________

**9.** ____________

**10.** ____________

**11.** ____________

**12.** ____________

Name Score

**Use translations to draw the graphs.**

1. Given the graph of $f(x) = |x|$ shown below, graph $g(x) = |x| - 5$.

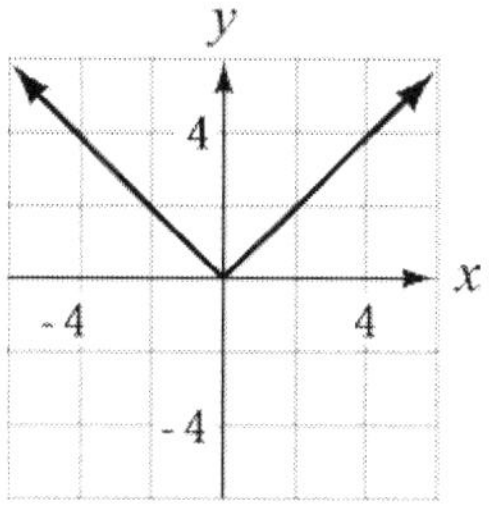

2. Given the graph of $f(x) = -\sqrt{x}$ shown below, graph $g(x) = -\sqrt{x-2} + 3$.

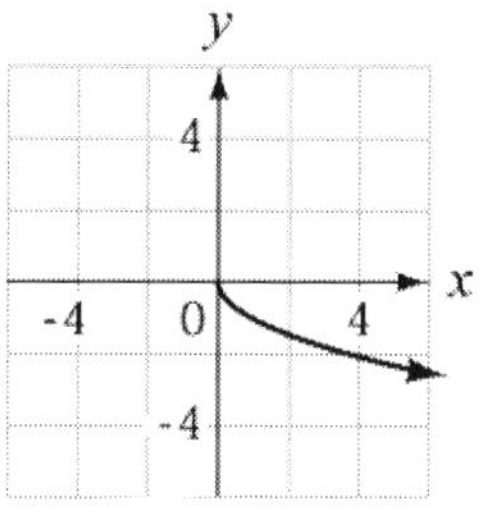

3. Given the graph of $f(x) = x^2$ shown below, graph $g(x) = (x-1)^2 - 2$.

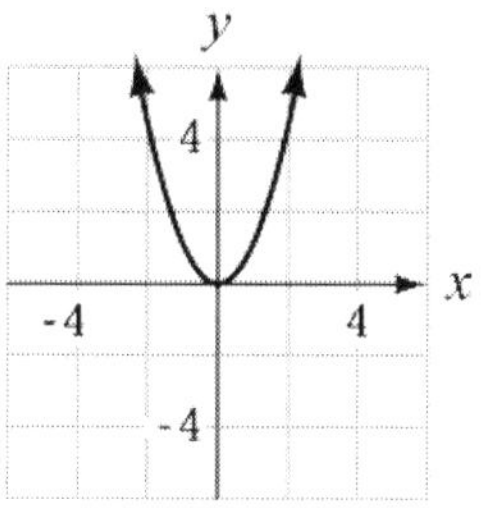

4. Given the graph of $f(x) = x^3$ shown below, graph $g(x) = (x+4)^3$.

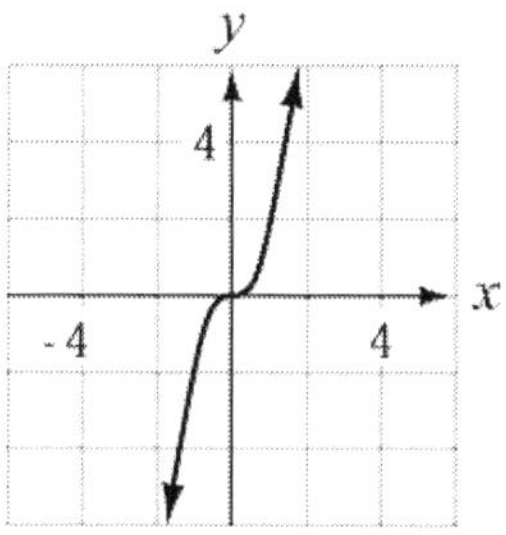

5. Given the graph of $f(x) = \sqrt{x}$ shown below, graph $g(x) = \sqrt{x+3}$.

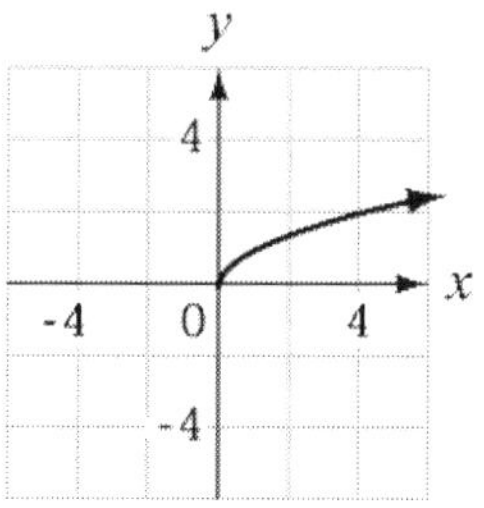

6. Given the graph of $f(x) = -|x|$ shown below, graph $g(x) = -|x-5|$.

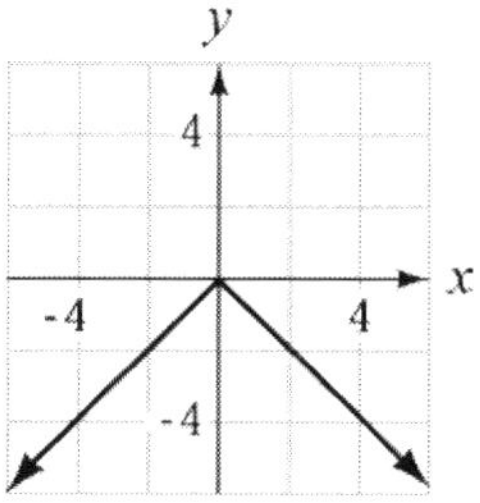

7. Given the graph of $f(x) = x^3$ shown below, graph $g(x) = (x-1)^3 + 4$.

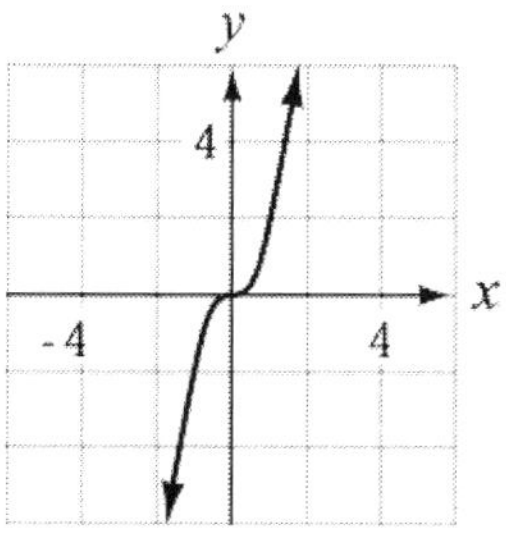

8. Given the graph of $f(x) = -x^2$ shown below, graph $g(x) = -x^2 + 3$.

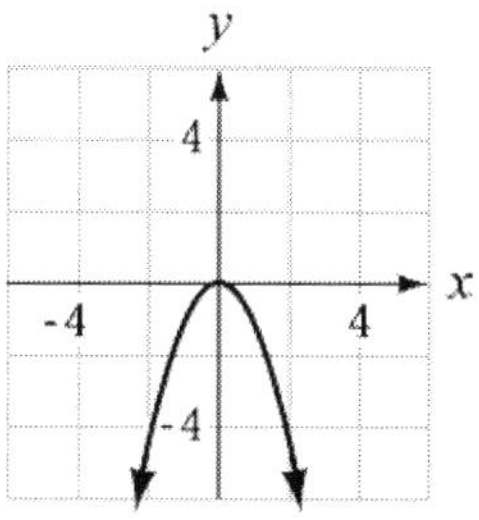

Name Score

**Given $f(x) = 3x^2 - 4$ and $g(x) = -5x + 6$, find**

1. $f(-2) - g(-2)$
2. $f(-1) + g(-1)$
3. $(f \cdot g)(-1)$
4. $\left(\frac{g}{f}\right)(-1)$
5. $\left(\frac{f}{g}\right)(-2)$
6. $g(5) - f(5)$

1. ______________
2. ______________
3. ______________
4. ______________
5. ______________
6. ______________

**Given $f(x) = 5x^2 - 4x - 6$ and $g(x) = 3x - 5$, find**

7. $f(5) - g(5)$
8. $g(1) - f(1)$
9. $f(2) + g(2)$
10. $(f \cdot g)(-2)$
11. $\left(\frac{f}{g}\right)\left(\frac{5}{3}\right)$
12. $\left(\frac{g}{f}\right)(0)$

7. ______________
8. ______________
9. ______________
10. ______________
11. ______________
12. ______________

**Given $f(x) = x^2 + 5x - 4$ and $g(x) = x^2 - 2x + 7$, find**

13. $f(-2) - g(-2)$
14. $f(3) + g(3)$
15. $(f \cdot g)(1)$
16. $\left(\frac{f}{g}\right)(-3)$

13. ______________
14. ______________
15. ______________
16. ______________

Name Score

**Given $f(x) = 5x - 4$ and $g(x) = 2x + 7$, find**

1. $f(g(1))$
2. $g(f(1))$
3. $f(g(2))$
4. $g(f(-2))$
5. $g(f(x))$
6. $f(g(x))$

1. ______
2. ______
3. ______
4. ______
5. ______
6. ______

**Given $h(x) = x^2 + 5$ and $g(x) = x - 3$, find**

7. $g(h(3))$
8. $h(g(x))$
9. $g(h(x))$
10. $h(g(0))$

7. ______
8. ______
9. ______
10. ______

**Given $h(x) = 2x^3 - 3$ and $f(x) = 4x - 5$, find**

11. $f(h(-1))$
12. $h(f(-1))$
13. $f(h(2))$
14. $f(h(x))$
15. $h(f(x))$
16. $f(h(0))$

11. ______
12. ______
13. ______
14. ______
15. ______
16. ______

Name Score

**Determine if the graph represents the graph of a 1-1 function.**

**1.**

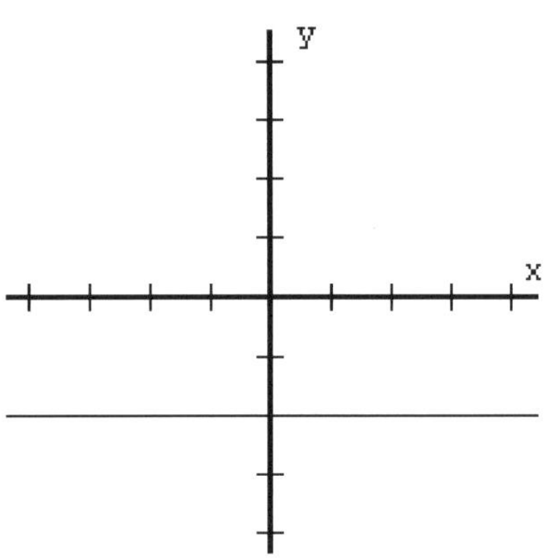


**2.**

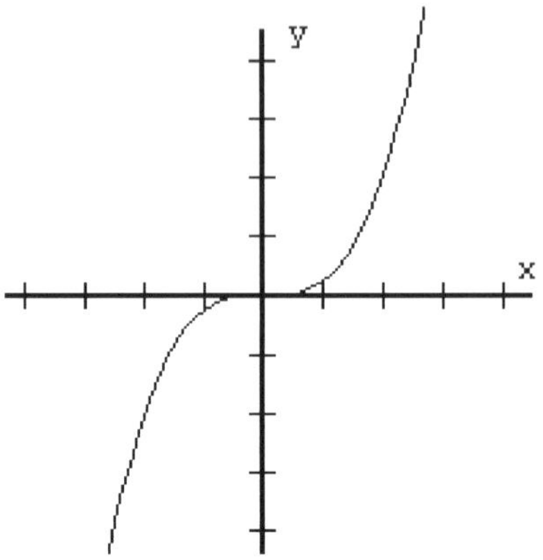


**3.**

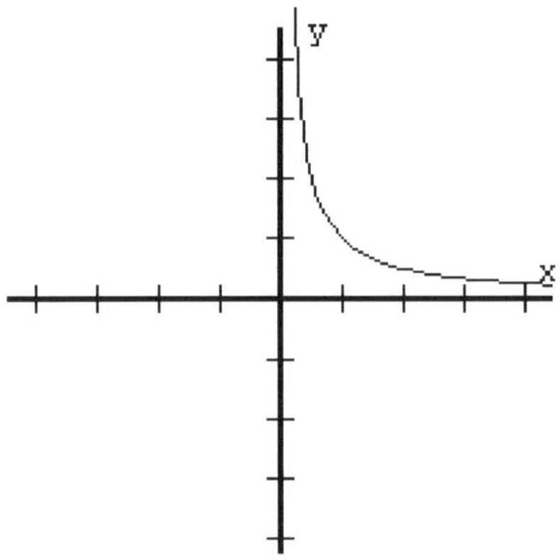


**4.**

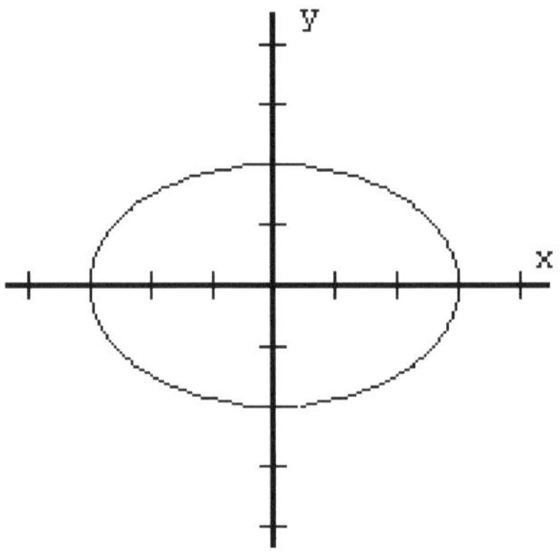


**5.**

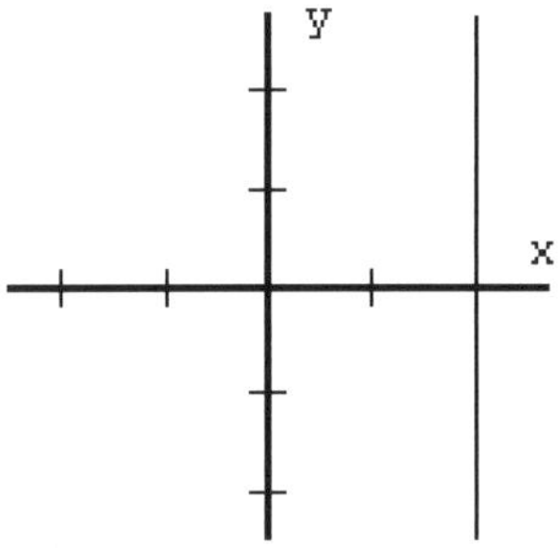


**6.**

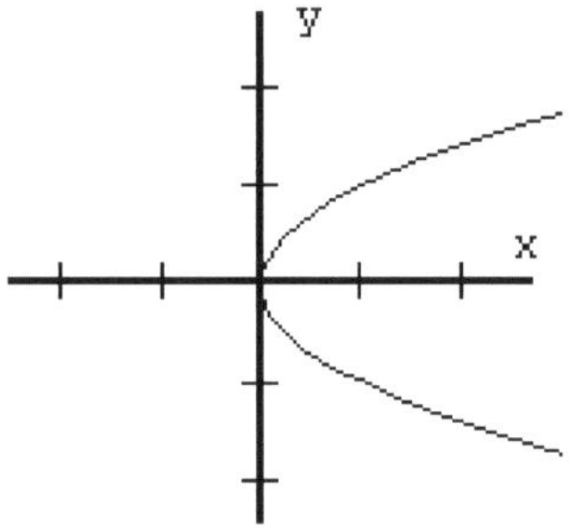


**7.**

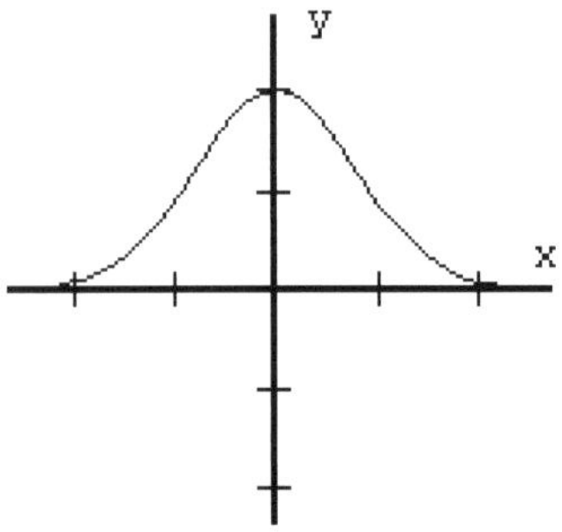


**8.**

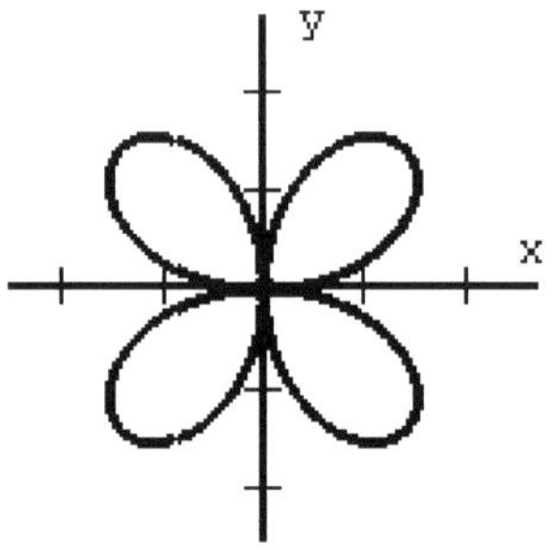


**1.** ________________

**2.** ________________

**3.** ________________

**4.** ________________

**5.** ________________

**6.** ________________

**7.** ________________

**8.** ________________

Name Score

**Find the inverse of the function.**

1. $f(x)=4x+8$

2. $f(x)=x+5$

3. $f(x)=3x-6$

4. $f(x)=-x+4$

5. $f(x)=-\frac{1}{4}x+2$

6. $f(x)=-2x+6$

7. $f(x)=\frac{2}{3}x-2$

8. $f(x)=\frac{1}{4}x-6$

9. $f(x)=3x+2$

10. $f(x)=5x+1$

11. $f(x)=-4x+2$

12. $f(x)=4x+3$

1. ______
2. ______
3. ______
4. ______
5. ______
6. ______
7. ______
8. ______
9. ______
10. ______
11. ______
12. ______

**Use the composition of Inverse Function Property to determine whether the functions are inverses.**

13. $f(x)=-\frac{1}{2}x,\ g(x)=-2x$

14. $f(x)=6x+5,\ g(x)=\frac{1}{6}x-\frac{5}{6}$

15. $g(x)=x+5,\ h(x)=5-x$

16. $h(x)=7x+3,\ g(x)=\frac{1}{7}x-3$

13. ______
14. ______
15. ______
16. ______

Name ______________________ Score ______________________

**Given the function $f(x) = 4^x$, find:**

1. $f(1)$
2. $f(-1)$
3. $f(0)$
4. $f(2)$
5. $f(-2)$
6. $f(3)$

1. ____________
2. ____________
3. ____________
4. ____________
5. ____________
6. ____________

**Given the function $f(x) = 3^{x+1}$, find:**

7. $f(0)$
8. $f(1)$
9. $f(-2)$

7. ____________
8. ____________
9. ____________

**Given the function $f(x) = \left(\frac{1}{3}\right)^{2x}$, find:**

10. $f(0)$
11. $f(1)$
12. $f(-1)$
13. $f\left(\frac{1}{2}\right)$
14. $f\left(-\frac{1}{2}\right)$
15. $f(2)$

10. ____________
11. ____________
12. ____________
13. ____________
14. ____________
15. ____________

**Given the function $f(x) = 3^{x^2}$, find:**

16. $f(0)$
17. $f(1)$
18. $f(2)$
19. $f(-1)$
20. $f(-2)$
21. $f(3)$

16. ____________
17. ____________
18. ____________
19. ____________
20. ____________
21. ____________

Name Score

**Graph.**

1. $f(x) = 2^x$

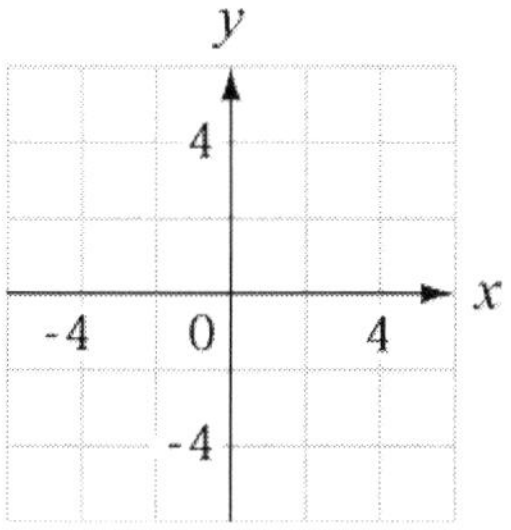


2. $f(x) = 2^x$

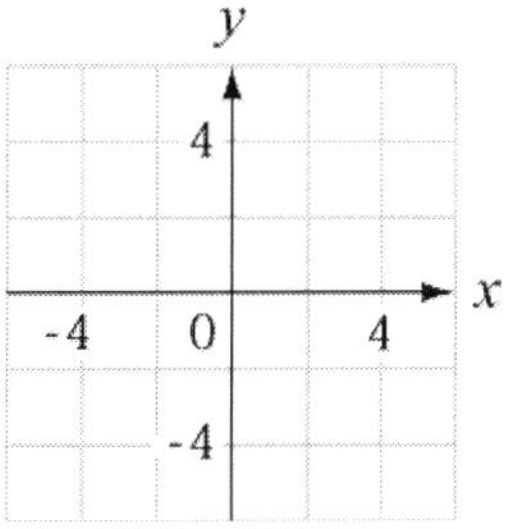


3. $f(x) = 3^{x+2}$

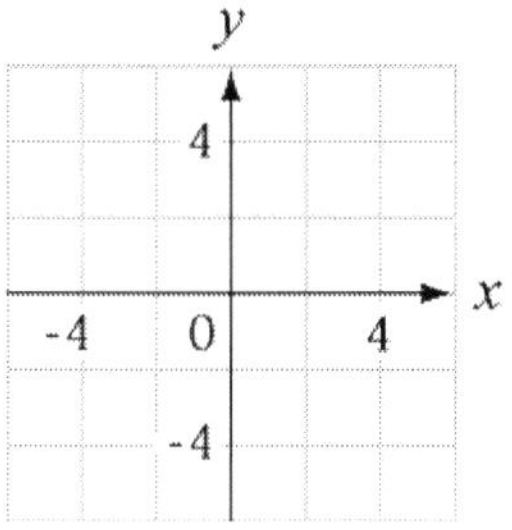


4. $f(x) = \left(\frac{1}{3}\right)^{-x-1}$

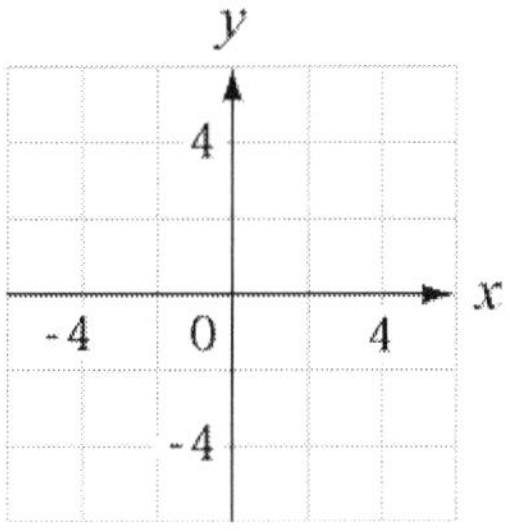


5. $f(x) = \left(\frac{1}{2}\right)^{x}$

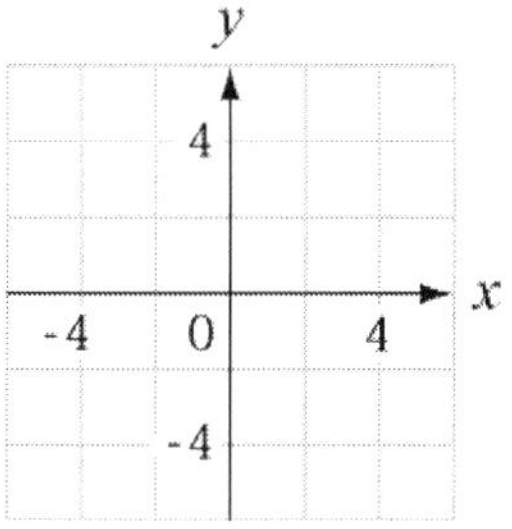


6. $f(x) = \left(\frac{1}{2}\right)^{x} - 1$

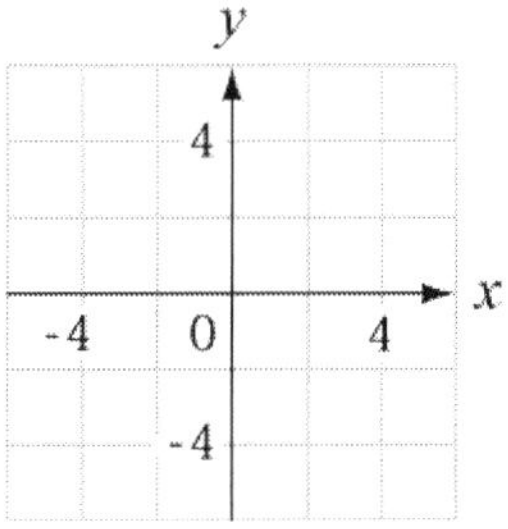


7. $f(x) = 2^x - 3$

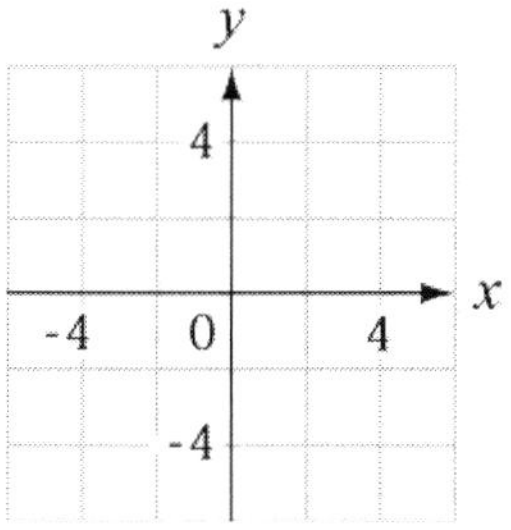


8. $f(x) = 2^x + 1$

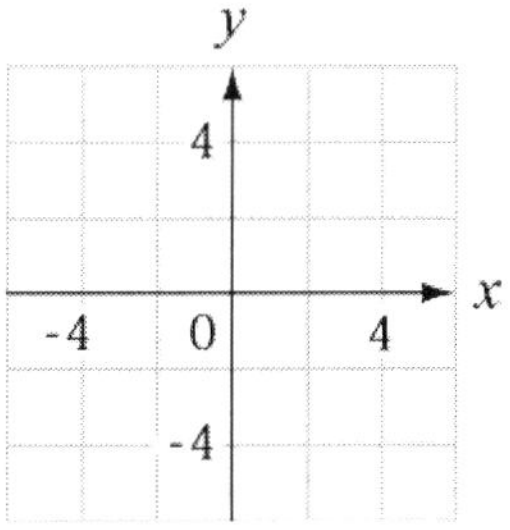

Name Score

**Write the exponential expression in logarithmic form.**

1. $4^3 = 64$
2. $7^2 = 49$
3. $64^{-2} = \frac{1}{8}$
4. $10^4 = 10{,}000$
5. $4^0 = 1$
6. $p^x = q$

1. ____________
2. ____________
3. ____________
4. ____________
5. ____________
6. ____________

**Write the logarithmic expression in exponential form.**

7. $\log_2 1 = 0$
8. $\log_2 \left(\frac{1}{4}\right) = -2$
9. $\ln x = c$

7. ____________
8. ____________
9. ____________

**Evaluate.**

10. $\log_5 25$
11. $\log 0.0001$
12. $\log_4 16$
13. $\ln e^4$
14. $\log_5 625$
15. $\log_8 1$

10. ____________
11. ____________
12. ____________
13. ____________
14. ____________
15. ____________

**Solve for *x*.**

16. $\log_8 x = 2$
17. $\log_4 x = 2$
18. $\log_6 x = 3$
19. $\log_7 x = 0$
20. $\log_{10} x = 4$
21. $\log_5 x = 3$

16. ____________
17. ____________
18. ____________
19. ____________
20. ____________
21. ____________

Name Score

**Express as a single logarithm with a coefficient of 1.**

**1.** $\log_2 x^2 + \log_2 y$

**2.** $\log_5 x^3 - \log_5 y^4$

**3.** $4\log_5 x$

**4.** $-3\log_6 x$

**5.** $3\log_6 x - 4\log_6 y$

**6.** $3\log_3 x - 2\log_3 y + 3\log_3 z$

**7.** $\frac{1}{3}(\log_6 x - \log_6 z)$

**8.** $5\log_3 x - 3(\log_3 y - \log_3 z)$

**9.** $\frac{1}{2}(5\log_5 x - 3\log_5 y + \log_5 z)$

1. ______
2. ______
3. ______
4. ______
5. ______
6. ______
7. ______
8. ______
9. ______

**Write the logarithm in expanded form.**

**10.** $\log_3(2x)$

**11.** $\log_2 x^3$

**12.** $\log_6\left(\frac{u^4}{v^5}\right)$

**13.** $\ln xy^3z^2$

**14.** $\log_6 \sqrt[3]{r^2 w}$

**15.** $\log_5 \sqrt{\frac{x^2}{y}}$

10. ______
11. ______
12. ______
13. ______
14. ______
15. ______

**Find the logarithm.**

**16.** $\log_5 8$

**17.** $\log_4 12$

**18.** $\log_6 30$

**19.** $\log_9 5$

**20.** $\log_3 2.9$

**21.** $\log_7 1.2$

16. ______
17. ______
18. ______
19. ______
20. ______
21. ______

Name Score

**Graph.**

**1.** $f(x) = \log_3 x$

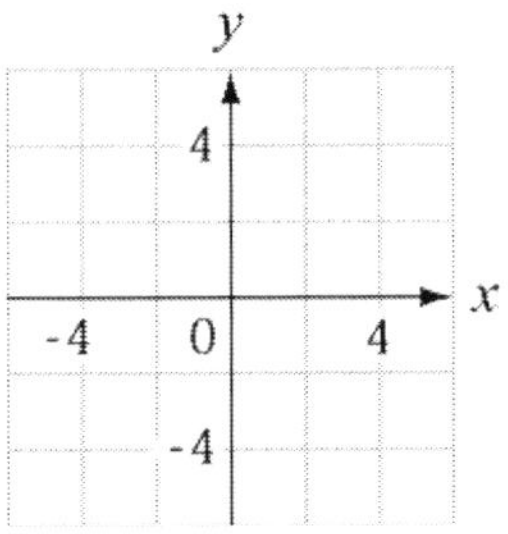


**2.** $f(x) = \log_2 x + 1$

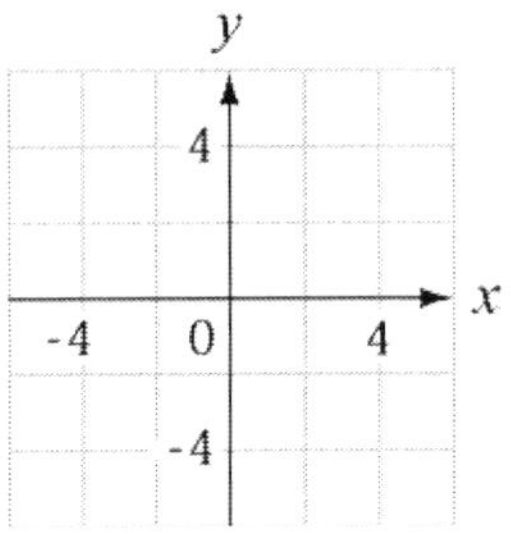


**3.** $f(x) = \log_2 3x$

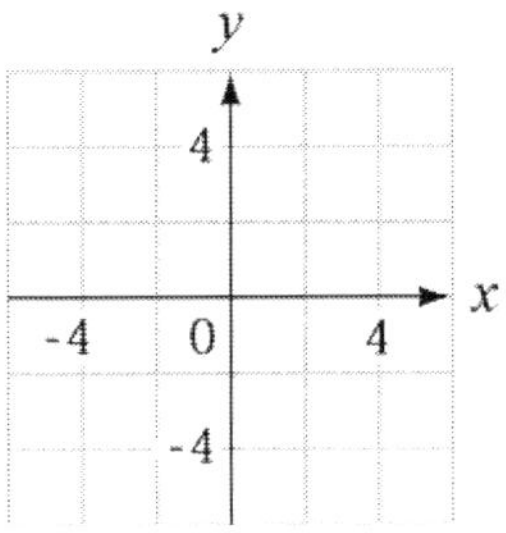


**4.** $f(x) = -\log_3(2 - x)$

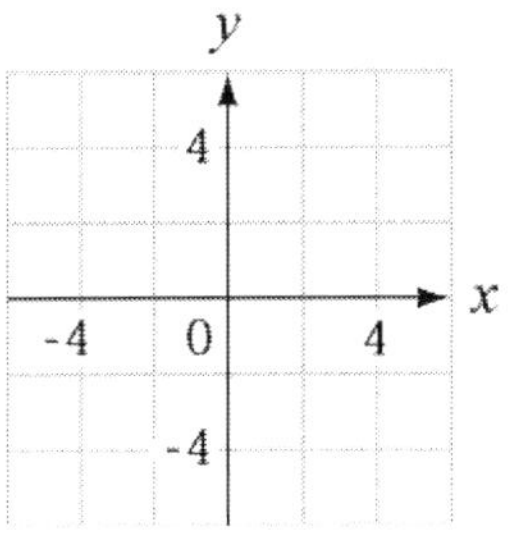


**5.** $f(x) = \log_3(x + 2)$

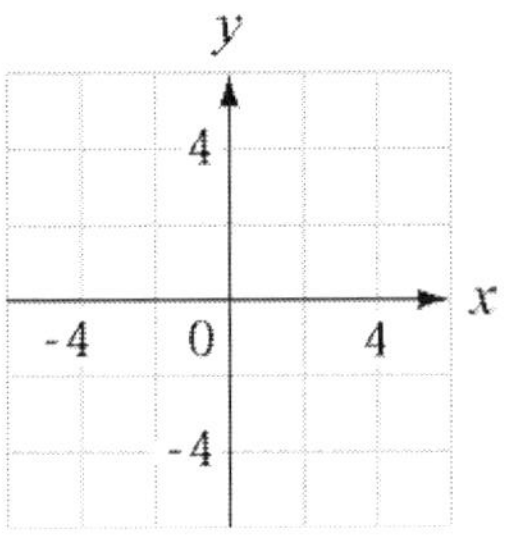


**6.** $f(x) = -\frac{1}{2}\log_2 x$

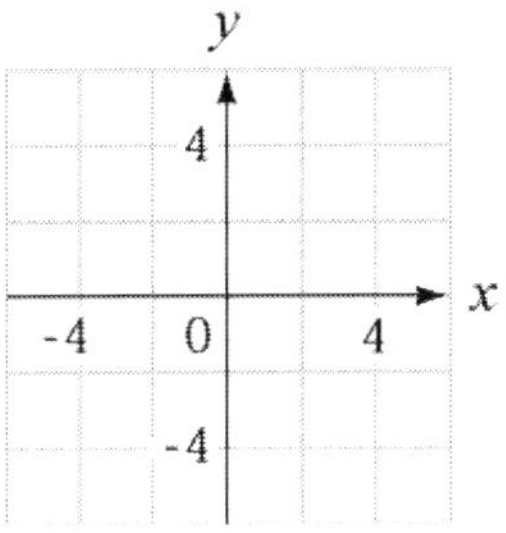


**7.** $f(x) = -\log_4 x$

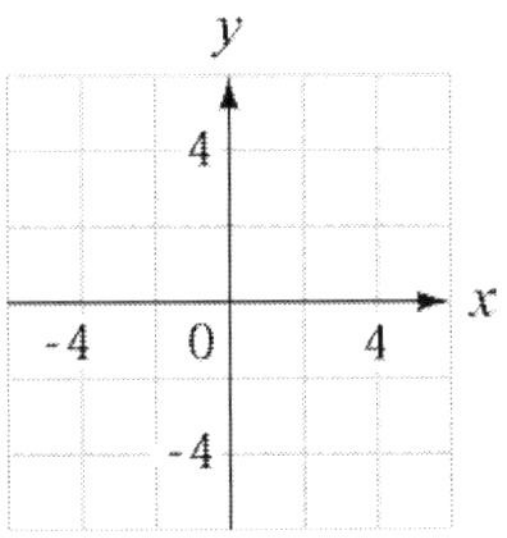


**8.** $f(x) = \log_4(x + 2)$

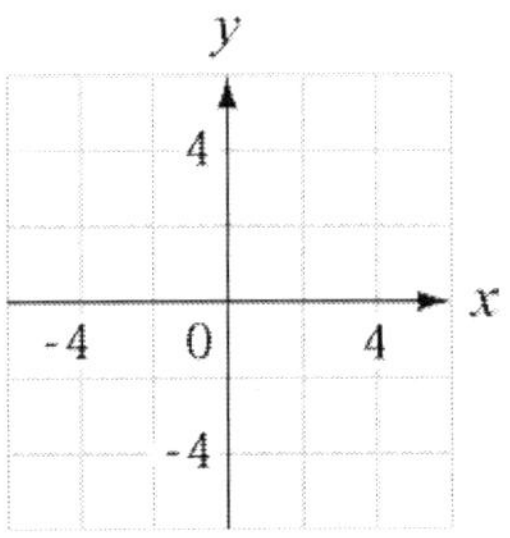

Name Score

**Solve for $x$. Round to the nearest ten-thousandth.**

1. $3^{2x-1} = 3^{x-3}$
2. $2^{4x-1} = 2^{x+2}$
3. $10^{3x-4} = 10^{x+6}$
4. $8^{x-3} = 8^{3x+5}$
5. $10^{x} = 6$
6. $12^{x} = 9$
7. $\left(\frac{1}{2}\right)^{x} = 4$
8. $\left(\frac{1}{3}\right)^{x} = 81$
9. $(1.4)^{x} = 3$
10. $(2.5)^{x} = 5$
11. $10^{x} = 24$
12. $10^{x} = 38$
13. $2^{-x} = 8$
14. $3^{-x} = 15$
15. $2^{x-1} = 9$
16. $3^{x+1} = 7$
17. $4^{2x-1} = 64$
18. $3^{-x+2} = 15$
19. $5^{x} = 25^{2x-1}$
20. $2^{-x+6} = 32^{x}$
21. $4^{x} = 2^{x+3}$
22. $6^{x-1} = 36^{x}$
23. $9^{x+6} = 27^{2x}$
24. $4^{3x} = 16^{x-2}$

1. ______
2. ______
3. ______
4. ______
5. ______
6. ______
7. ______
8. ______
9. ______
10. ______
11. ______
12. ______
13. ______
14. ______
15. ______
16. ______
17. ______
18. ______
19. ______
20. ______
21. ______
22. ______
23. ______
24. ______

Name Score

**Solve for *x*.**

1. $\log_5(x+1)=2$

2. $\log_7(x+2)=1$

3. $\log_3(2x-1)=3$

4. $\log_5(x-1)=2$

5. $\log_4(3x+1)=2$

6. $\log_3(2x-3)=4$

7. $\log_5\left(x^2-4x\right)=1$

8. $\log_4\left(x^2+6x\right)=2$

9. $\log_6\left(x^2-5x\right)=2$

10. $\log_3 x=\log_3(2-x)$

11. $\log_5\left(\frac{3x}{x-1}\right)=1$

12. $\log_4 x+\log_4(3x-4)=\log_4 15$

13. $\log_7 5x=\log_7 2+\log_7(x-3)$

14. $\log_5 9x-\log_5\left(x^2-2\right)=\log_5 6$

15. $\log_6 3x=\log_6 2+\log_6\left(x^2-1\right)$

16. $\log_4 7x=\log_4 3+\log_4\left(x^2-2\right)$

1. ____________
2. ____________
3. ____________
4. ____________
5. ____________
6. ____________
7. ____________
8. ____________
9. ____________
10. ____________
11. ____________
12. ____________
13. ____________
14. ____________
15. ____________
16. ____________

Name Score

**Use the compound interest formula $P = A(1+i)^n$, where $A$ is the original value of an investment, $i$ is the interest rate per compounding period, $n$ is the total number of compounding periods, and $P$ is the value of the investment after $n$ periods.**

1. An investor deposits \$8000 into an account that earns 12% interest compounded semi-annually. What is the value of the investment after 4 years?

2. An investor deposits \$5000 into an account that earns 12% annual interest compounded semi-annually. In approximately how many years will the investment be worth twice the original amount?

1. ____________

2. ____________

**The percent of correct welds a student can make will increase with practice and can be approximated by the equation $P = 100[1 - (0.75)^t]$, where $P$ is the percent of correct welds and $t$ is the number of weeks of practice.**

3. Find the percent of correct welds a student will make after four weeks of practice.

4. How many weeks of practice are necessary before a student will make 95% of the welds correctly?

3. ____________

4. ____________

**The percent of light that will pass through a material is given by the equation $\log P = -kd$, where $P$ is the percent of light passing through the material, $k$ is a constant that depends on the material, and $d$ is the thickness of the material in centimeters.**

5. The constant $k$ for a piece of tinted glass is 0.5. How thick a piece of this glass is needed so that 60% of the light that is incident to the glass will pass through it?

6. The constant $k$ for a piece of opaque glass that is 0.25 cm thick is 0.5. Find the percent of light that will pass through the glass.

5. ____________

6. ____________

**Use the Richter equation $M = \log\frac{I}{I_0}$, where $M$ is the magnitude of an earthquake, $I$ is the intensity of its shock waves, and $I_0$ is a constant.**

7. How many times stronger is an earthquake that has magnitude 5.8 on the Richter scale than one that has magnitude 3.2 on the scale?

8. How many times stronger is an earthquake that has magnitude 6.4 on the Richter scale than one that has magnitude 1.6 on the scale?

7. ____________

8. ____________

Name Score

**Find the vertex and axis of symmetry of the parabola. Then sketch its graph.**

**1.** $y = x^2 - x - 2$

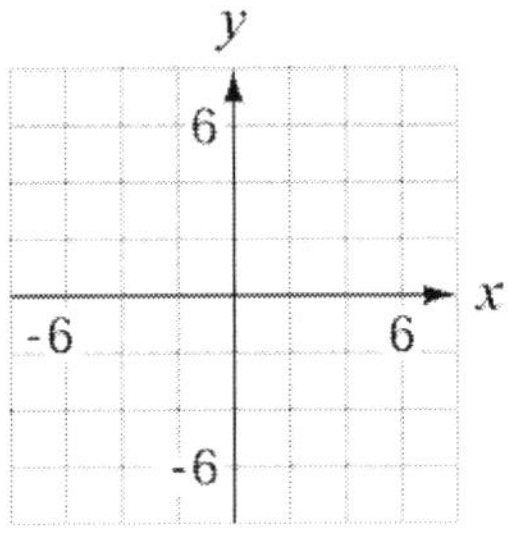


**2.** $y = x^2 + 2x - 3$

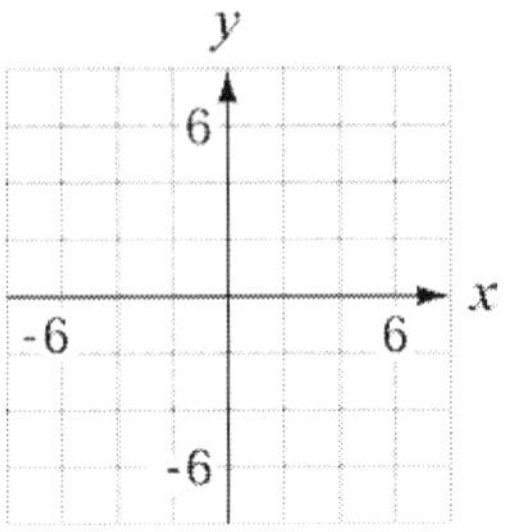


**3.** $y = -x^2 + 2x + 4$

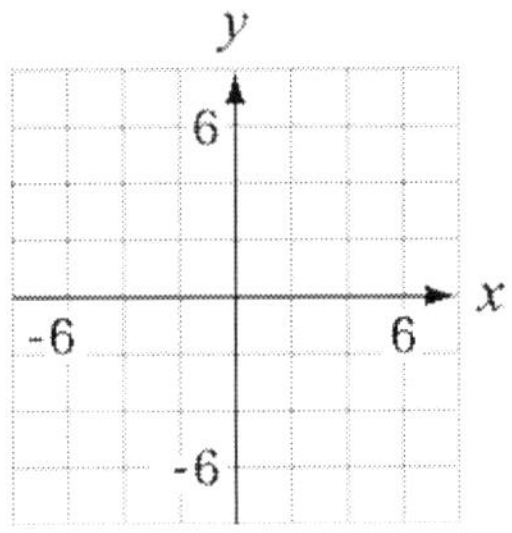


**4.** $y = -x^2 + 6x - 5$

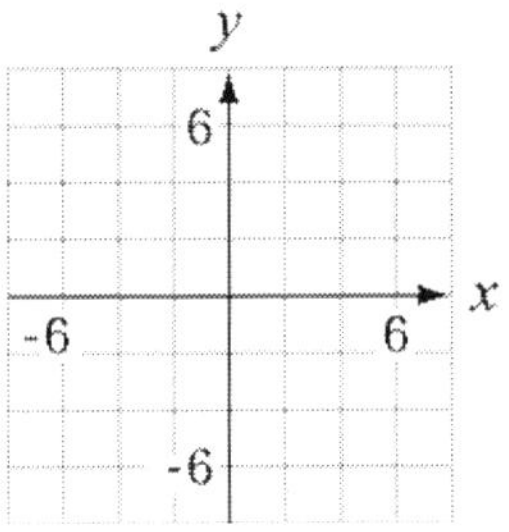


**5.** $x = y^2 + 2x + 3$

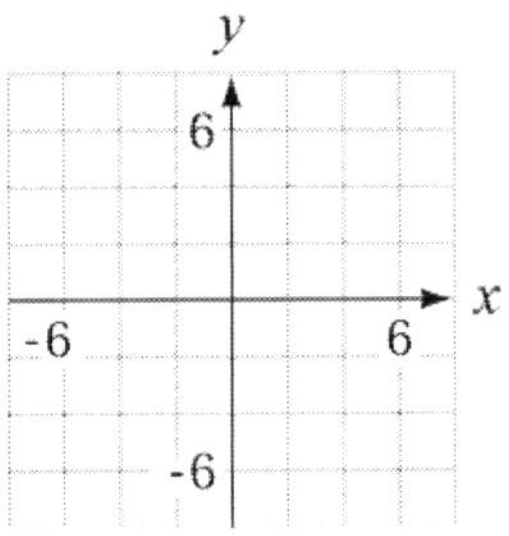


**6.** $x = 3y^2 - 3$

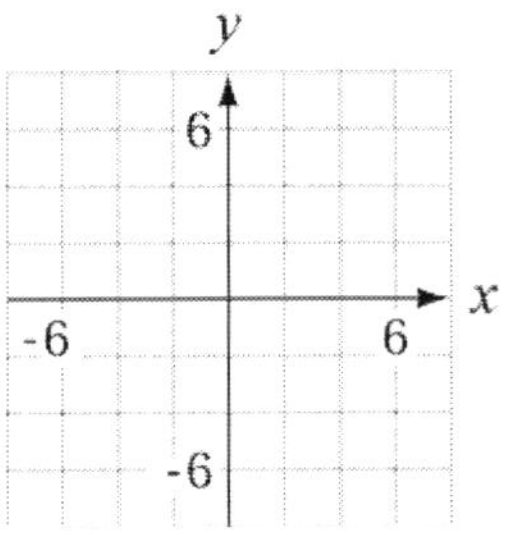


**7.** $x = -\frac{1}{4}y^2 + 4$

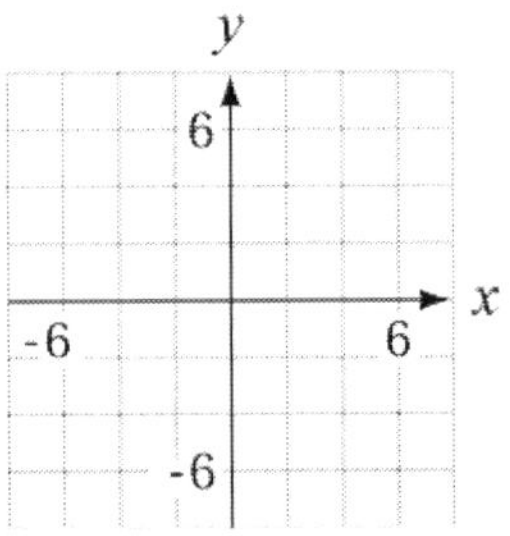


**8.** $x = -\frac{1}{2}y^2 + 2$

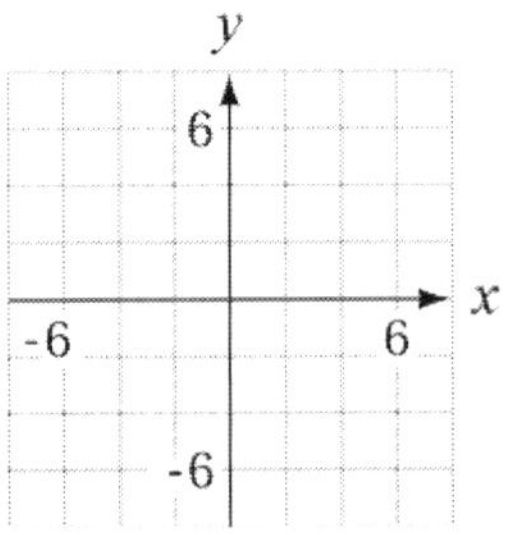


**1.** ____________

**2.** ____________

**3.** ____________

**4.** ____________

**5.** ____________

**6.** ____________

**7.** ____________

**8.** ____________

Name | Score

**Sketch a graph of the circle.**

1. $(x-1)^2+(y+1)^2=4$

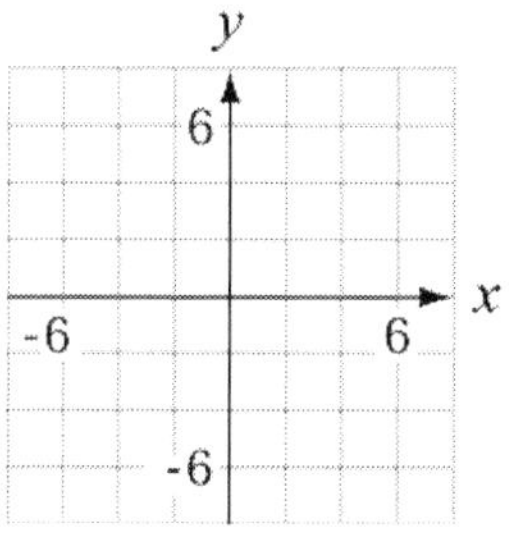


2. $(x+1)^2+(y-2)^2=9$

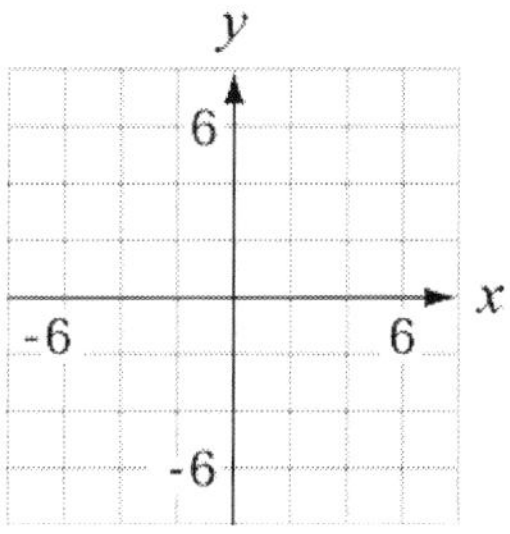


3. $(x-1)^2+(y+3)^3=9$

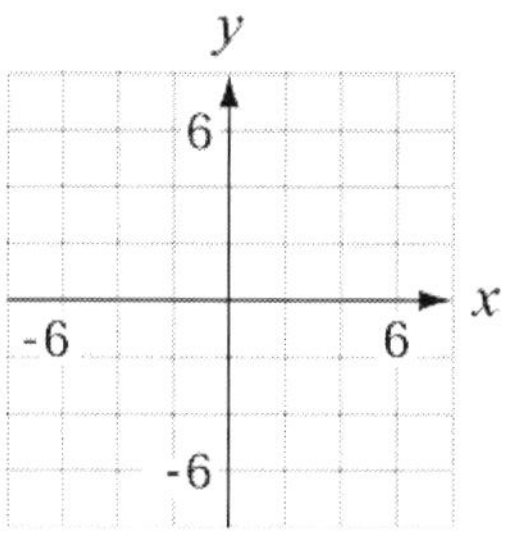


4. $(x-3)^2+(y+2)^2=16$

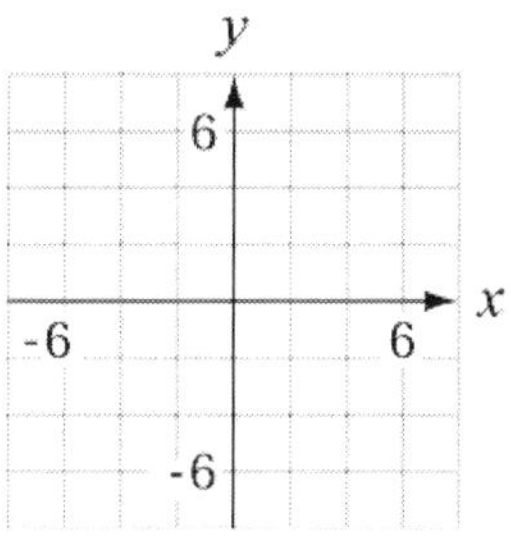


5. Find the equation of the circle with radius 4 and center (0, –4). Then sketch the graph.

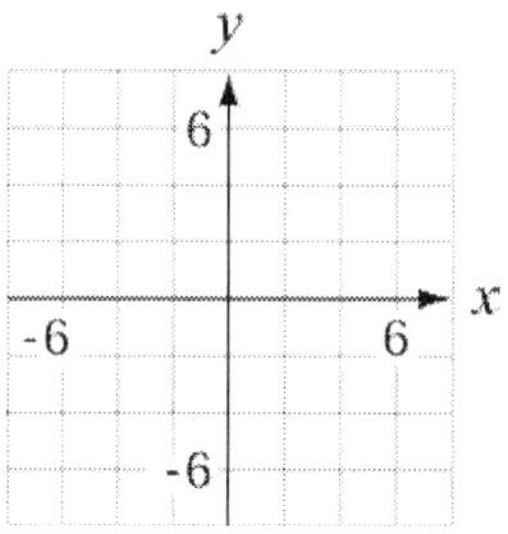


6. Find the equation of the circle with radius 2 and center (–2, –1). Then sketch the graph.

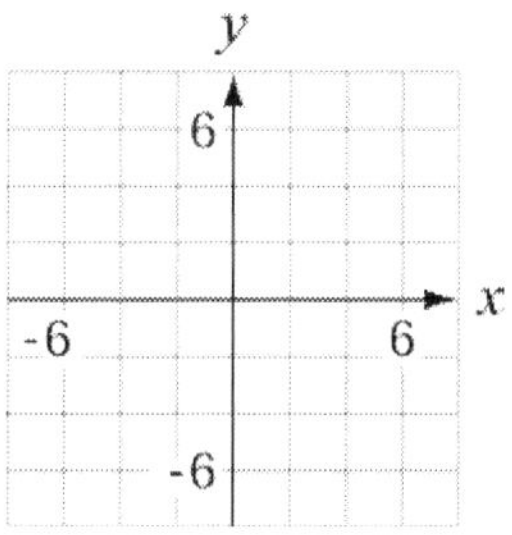


7. Find the equation of the circle that passes through point (3, 0) and whose center is (–1, 3). Then sketch its graph.

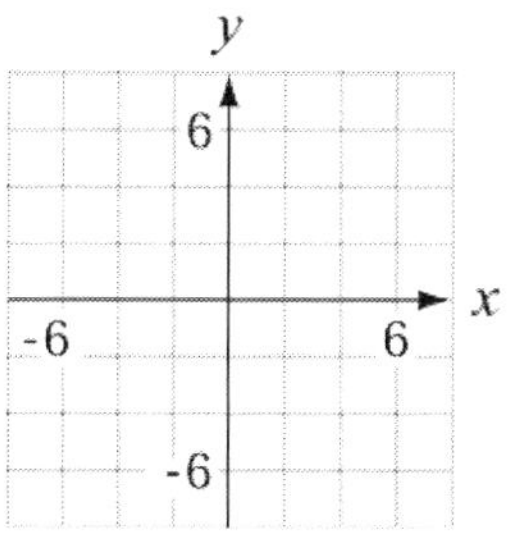


8. Find the equation of the circle that passes through point (–2, –3) and whose center is (2, –3). Then sketch its graph.

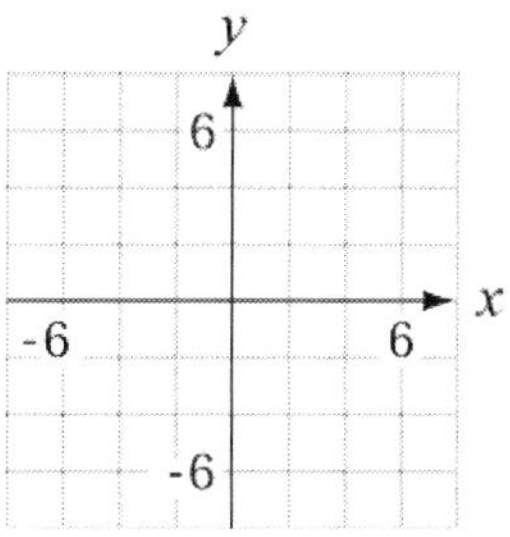

Name Score

**Write the equation of the circle in standard form. Then sketch its graph.**

**1.** $x^2 - 10x + y^2 + 2y + 22 = 0$

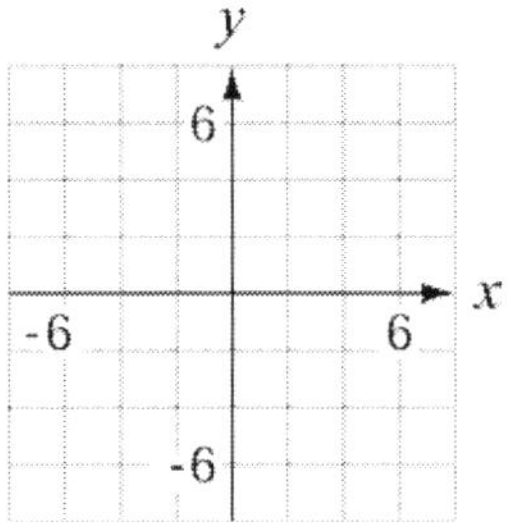


**2.** $x^2 + y^2 + 8x - 4y + 4 = 0$

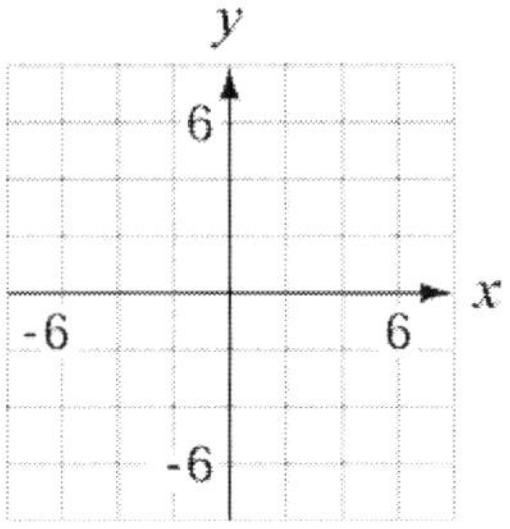


**3.** $x^2 + y^2 - 10x + 5y + 25 = 0$

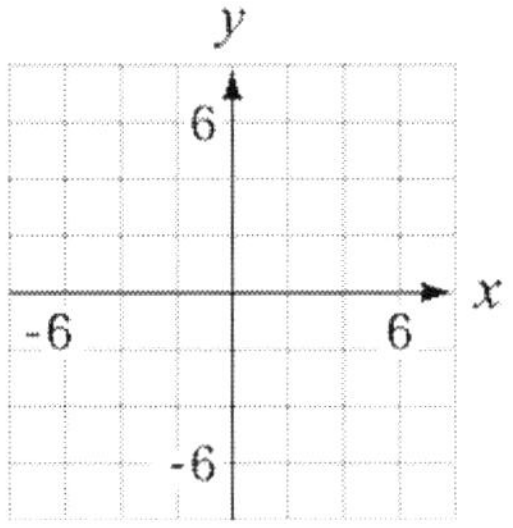


**4.** $x^2 + y^2 + 6x - 10y + 9 = 0$

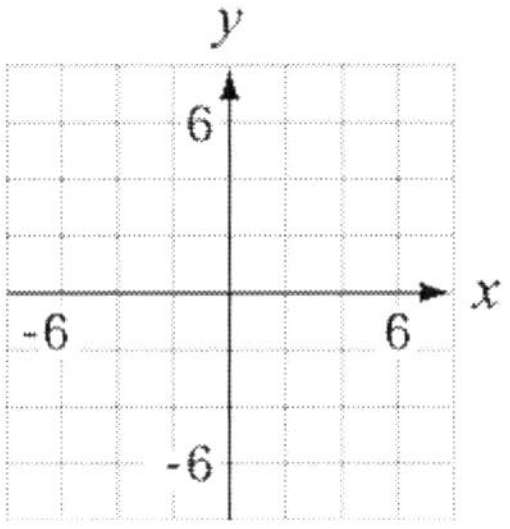


**5.** $x^2 + y^2 - 6x + 8y - 16 = 0$

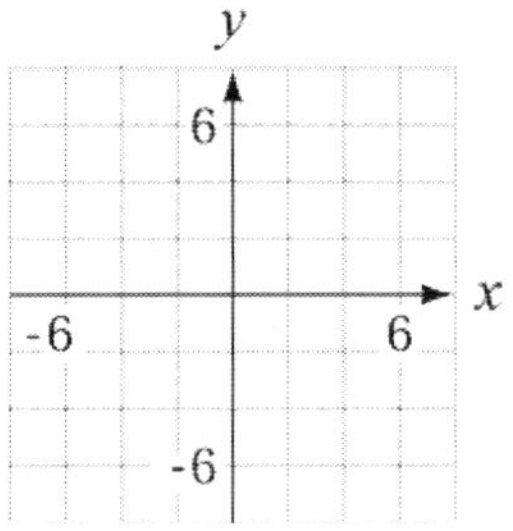


**6.** $x^2 + y^2 - 6x - 4y + 9 = 0$

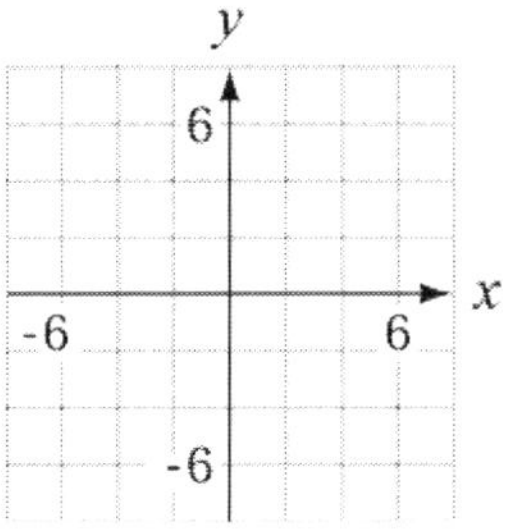


**7.** $x^2 + y^2 + 6x - 4y - 3 = 0$

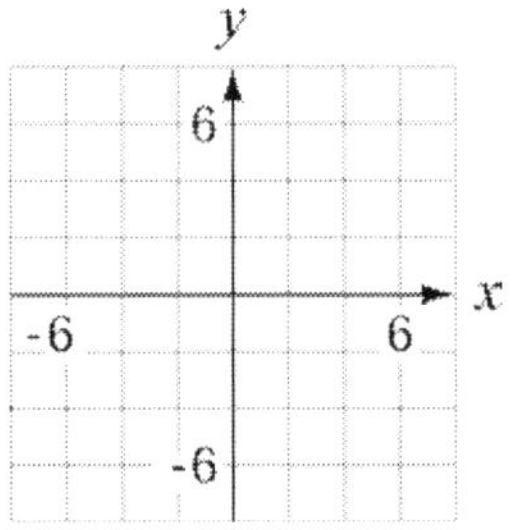


**8.** $x^2 + y^2 + 6x - 2y + 6 = 0$

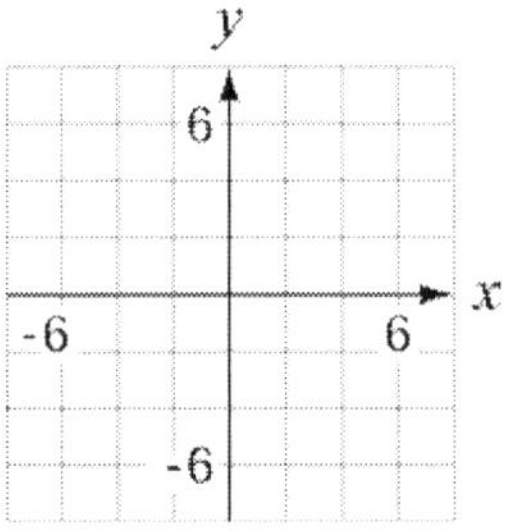


**1.** ______________

**2.** ______________

**3.** ______________

**4.** ______________

**5.** ______________

**6.** ______________

**7.** ______________

**8.** ______________

Name Score

**Sketch a graph of the ellipse.**

**1.** $\frac{x^2}{9}+\frac{y^2}{4}=1$

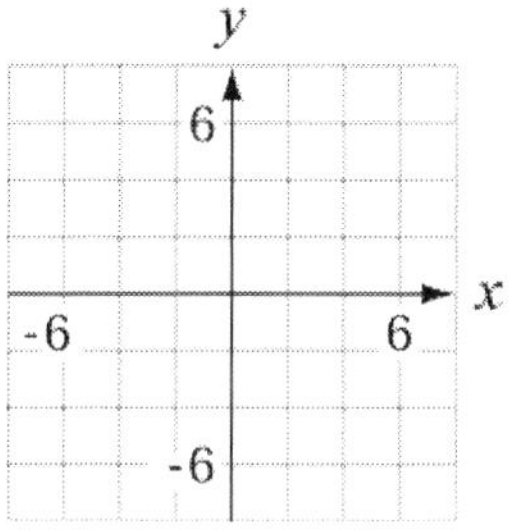


**2.** $\frac{x^2}{25}+\frac{y^2}{9}=1$

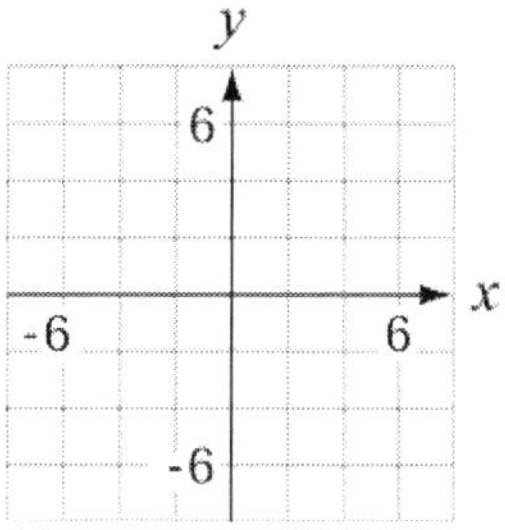


**3.** $\frac{x^2}{36}+\frac{y^2}{9}=1$

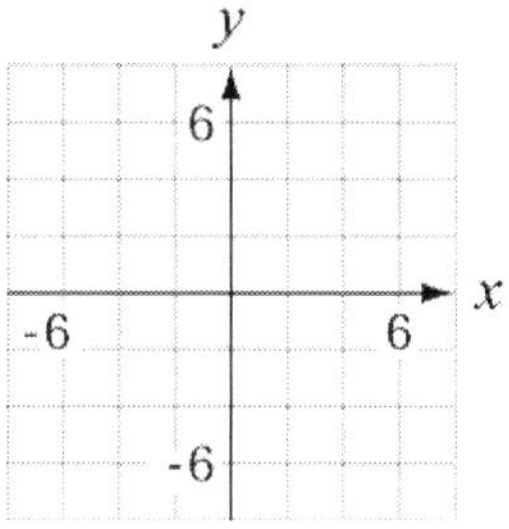


**4.** $\frac{x^2}{4}+\frac{y^2}{25}=1$

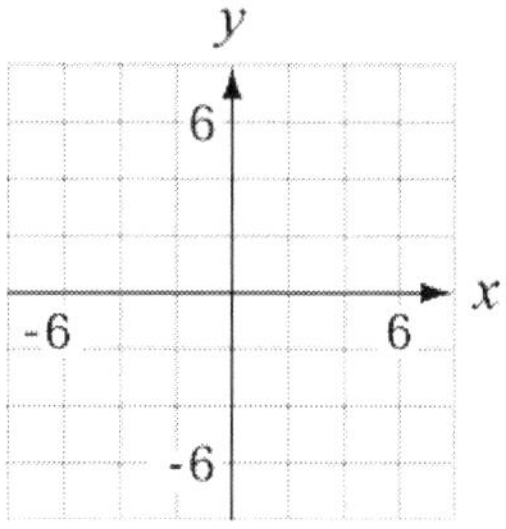


**5.** $\frac{x^2}{49}+\frac{y^2}{25}=1$

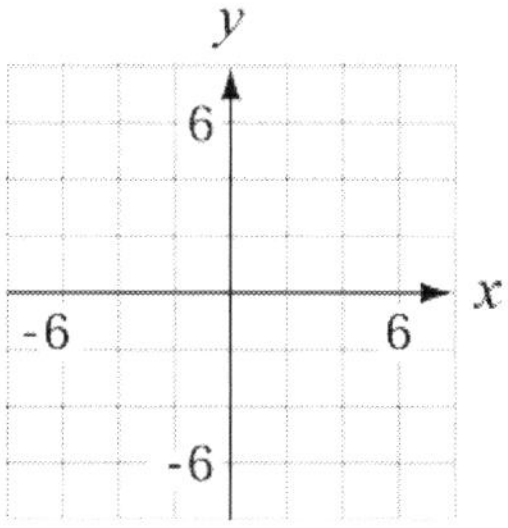


**6.** $\frac{x^2}{49}+\frac{y^2}{4}=1$

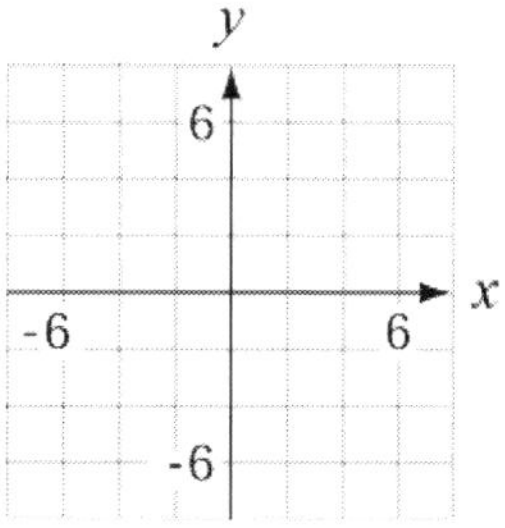


**7.** $\frac{x^2}{25}+\frac{y^2}{49}=1$

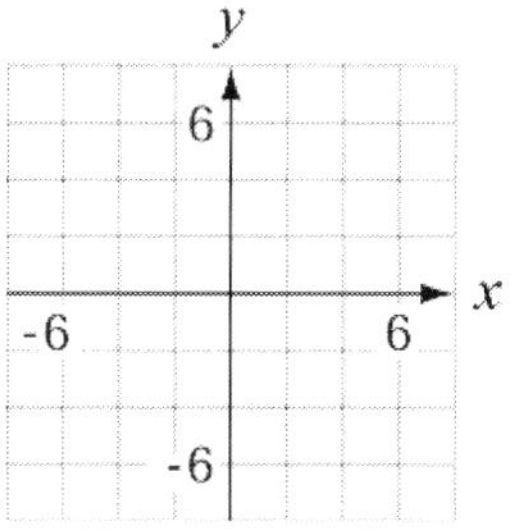


**8.** $\frac{x^2}{36}+\frac{y^2}{25}=1$

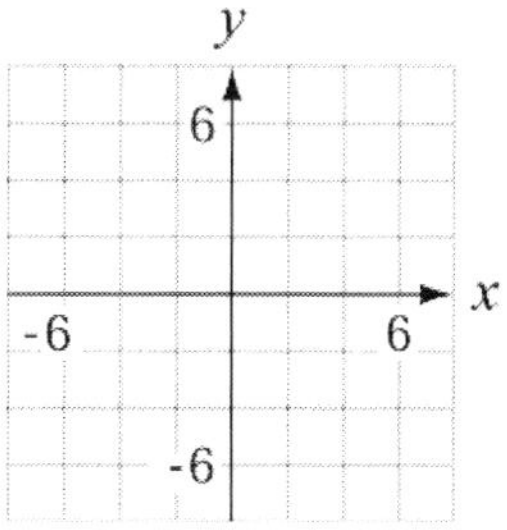

Name Score

**Sketch a graph of the hyperbola.**

**1.** $\frac{y^2}{4} - \frac{x^2}{9} = 1$

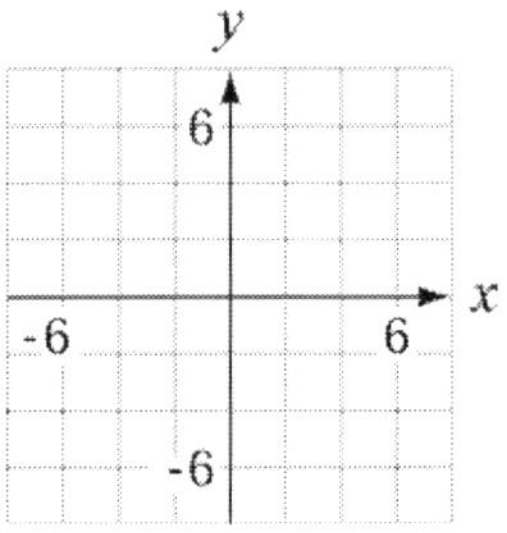


**2.** $\frac{x^2}{9} - \frac{y^2}{25} = 1$

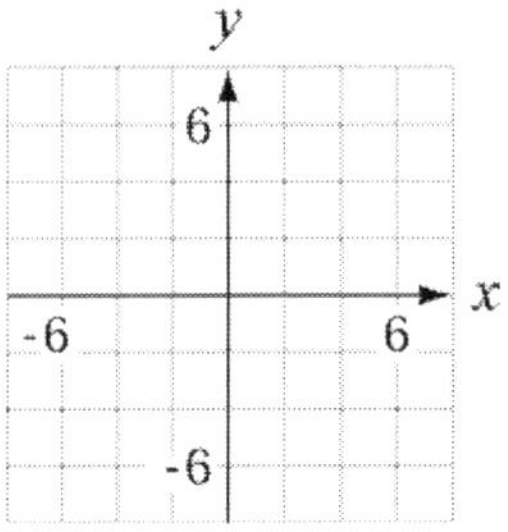


**3.** $\frac{y^2}{49} - \frac{x^2}{4} = 1$

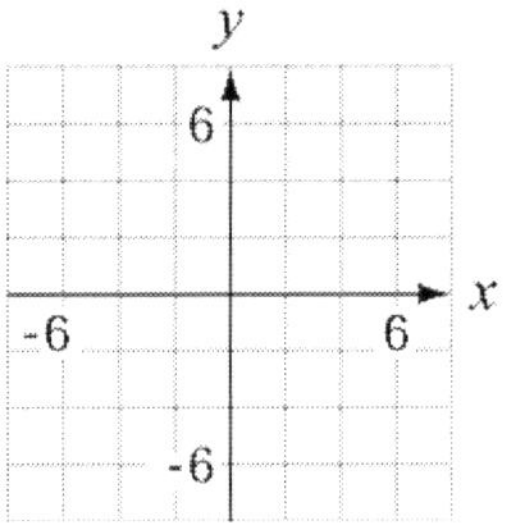


**4.** $\frac{y^2}{9} - \frac{x^2}{9} = 1$

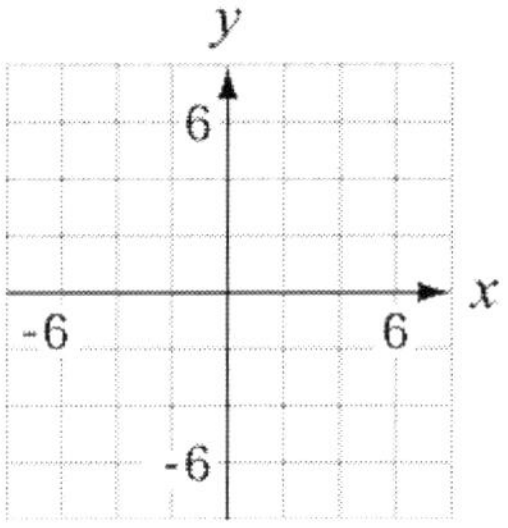


**5.** $\frac{x^2}{4} - \frac{y^2}{25} = 1$

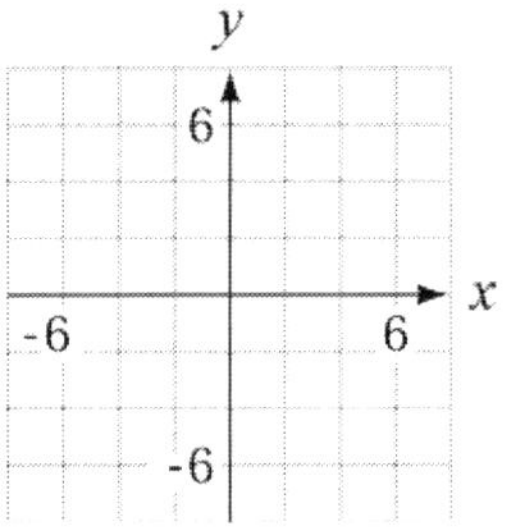


**6.** $\frac{x^2}{49} - \frac{y^2}{9} = 1$

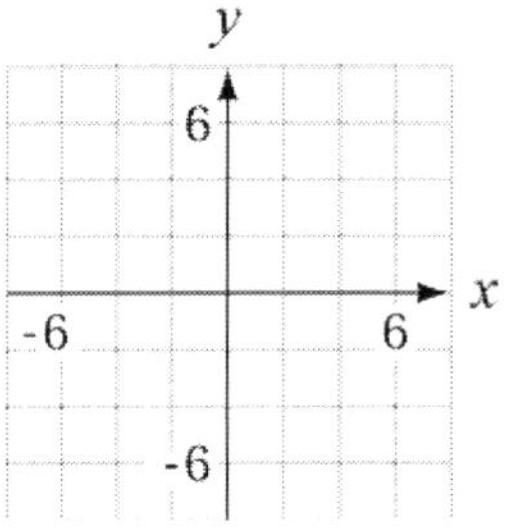


**7.** $\frac{y^2}{36} - \frac{x^2}{9} = 1$

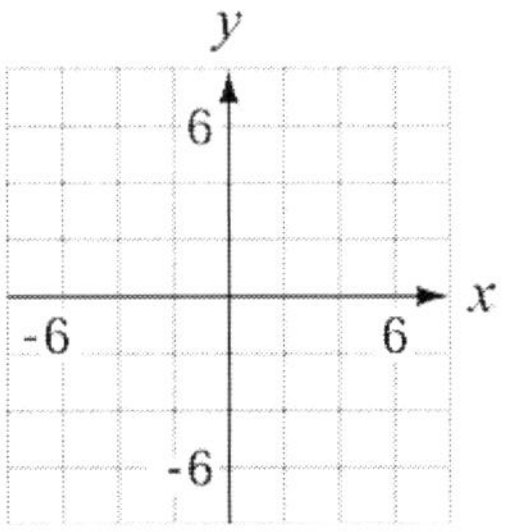


**8.** $\frac{y^2}{9} - \frac{x^2}{36} = 1$

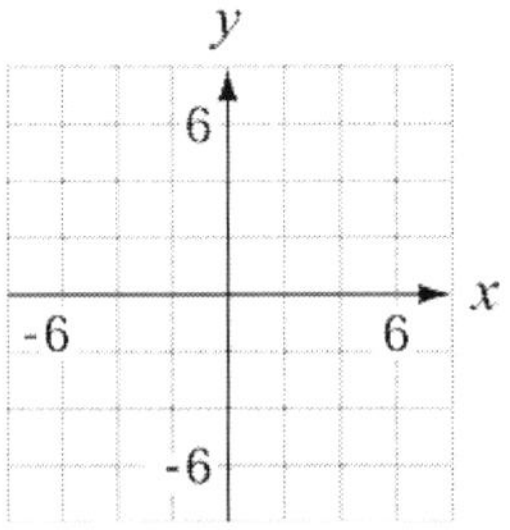

Name Score

**Solve.**

1. $y = x^2 - 3x - 4$
   $y = 2x + 2$

2. $y = x^2 - 3x - 9$
   $y = 2x + 5$

3. $y = x^2 + 4x - 18$
   $2x - y = 3$

4. $y = x^2 - 5x$
   $y = 2x$

5. $y^2 = x + 2$
   $y = x$

6. $x^2 + 4y^2 = 13$
   $x + 2y = -5$

7. $x^2 + y^2 = 10$
   $x + y = 4$

8. $x^2 + y^2 = 25$
   $2x - y = 2$

9. $x^2 + y^2 = 13$
   $x - y = 1$

10. $x^2 - y^2 = 7$
    $x - 2y = -2$

11. $y^2 - x^2 = 5$
    $x + y = 1$

12. $x^2 + y^2 = 45$
    $x^2 - y^2 = 27$

13. $2x^2 + y^2 = 8$
    $x^2 - 2y^2 = 4$

14. $x^2 + 4y^2 = 25$
    $x^2 - y^2 = 5$

15. $2x^2 + y^2 = 6$
    $x - 2y = -5$

16. $x^2 + 4y^2 = 17$
    $3x^2 - y^2 = -1$

1. ______________
2. ______________
3. ______________
4. ______________
5. ______________
6. ______________
7. ______________
8. ______________
9. ______________
10. ______________
11. ______________
12. ______________
13. ______________
14. ______________
15. ______________
16. ______________

Name Score

**Graph the solution set.**

**1.** $y < x^2 - 5x + 4$

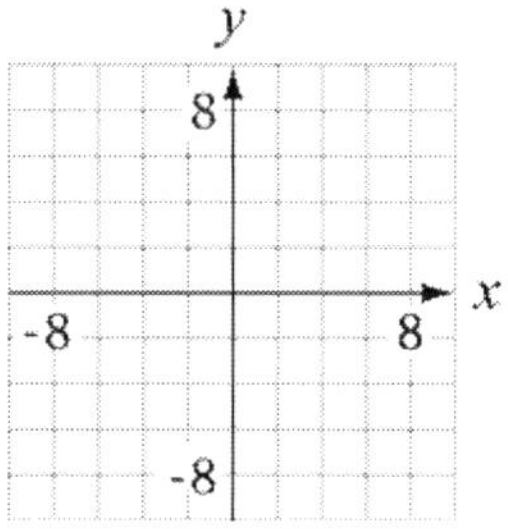


**2.** $y > x^2 - 2x$

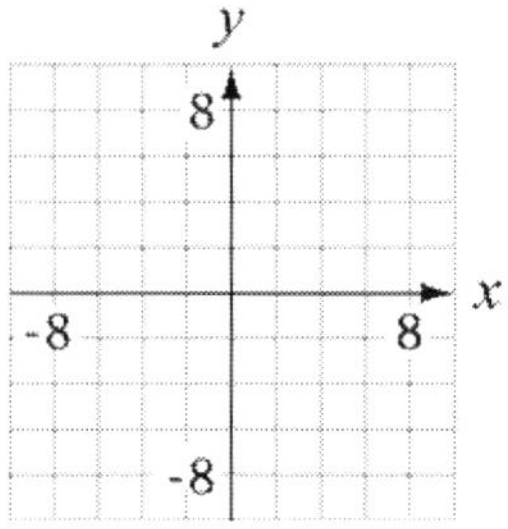


**3.** $(x-2)^2 + (y+1)^2 \le 4$

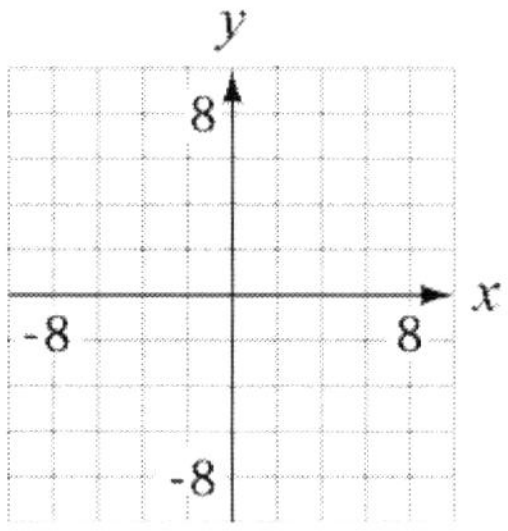


**4.** $(x+1)^2 + (y-3)^2 > 4$

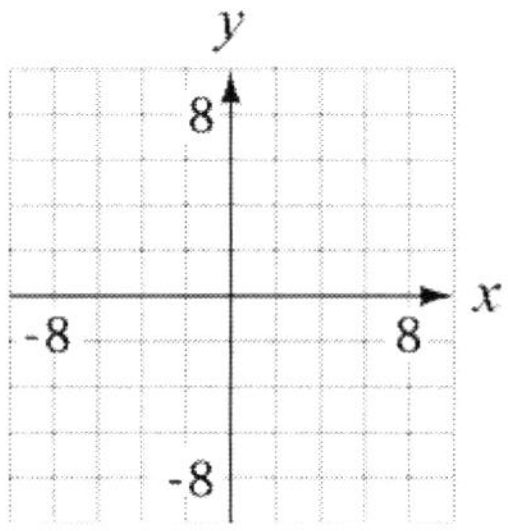


**5.** $\frac{x^2}{9} + \frac{y^2}{16} > 1$

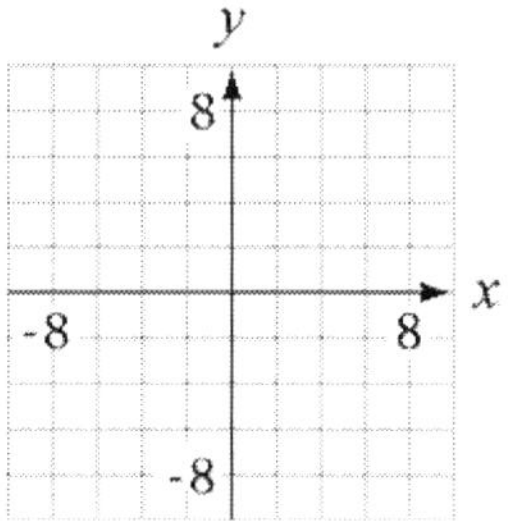


**6.** $\frac{x^2}{36} + \frac{y^2}{9} \le 1$

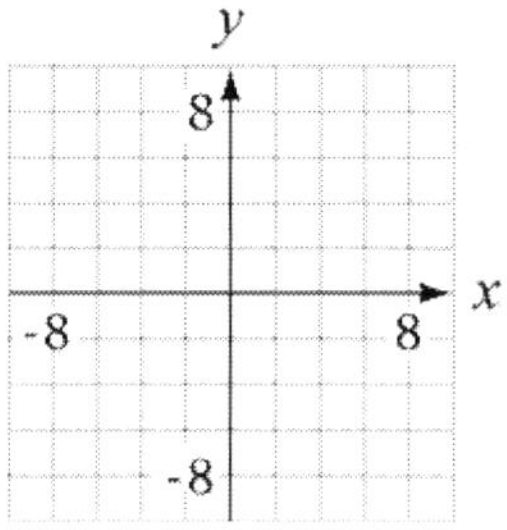


**7.** $\frac{x^2}{25} - \frac{y^2}{9} \ge 1$

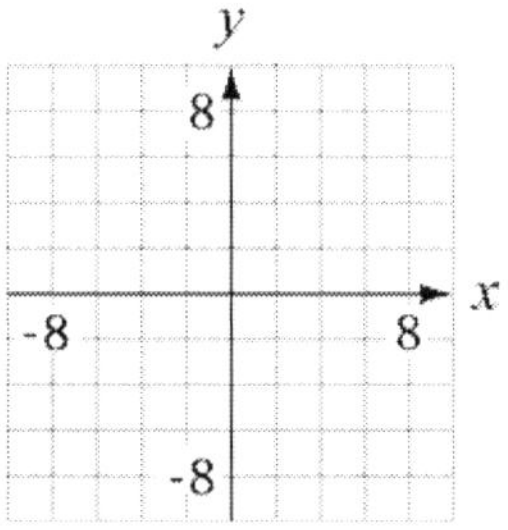


**8.** $\frac{y^2}{4} - \frac{x^2}{9} \le 1$

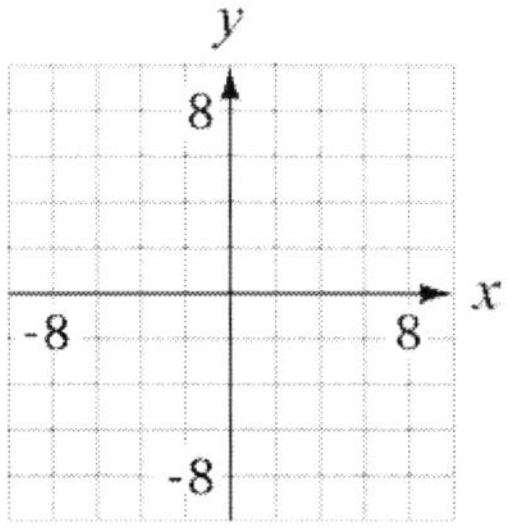

Name Score

**Graph the solution set.**

**1.** $y \le (x-1)^2$
$x + y > 3$

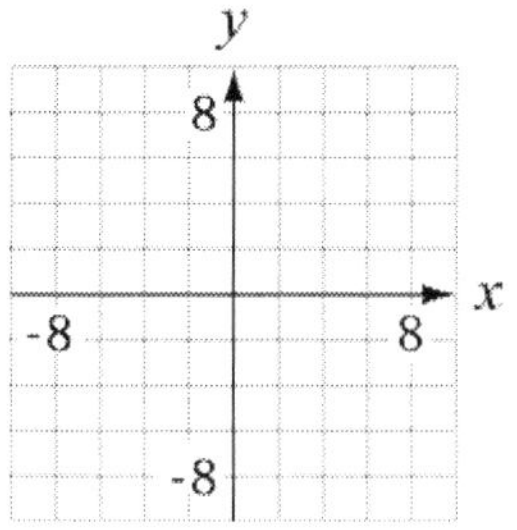


**2.** $\dfrac{x^2}{25} + \dfrac{y^2}{4} < 1$
$x + y \ge 8$

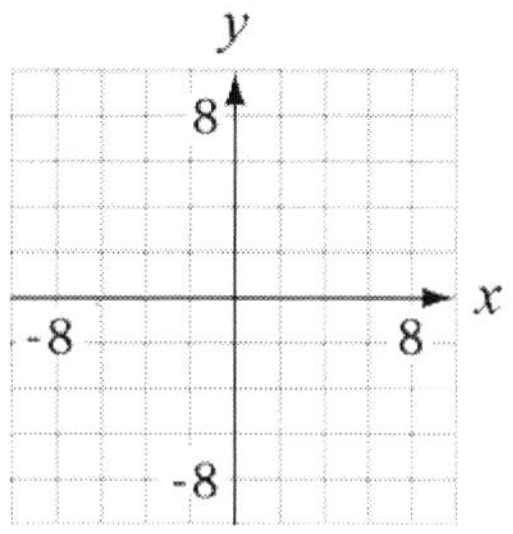


**3.** $y > x^2 - 8$
$y < x - 4$

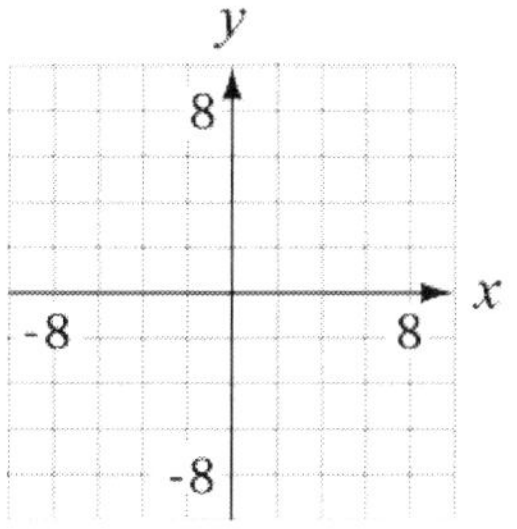


**4.** $\dfrac{x^2}{36} + \dfrac{y^2}{16} \le 1$
$x + 2y \le 2$

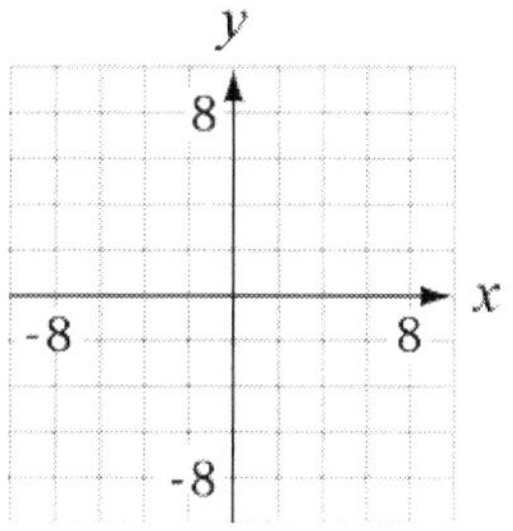


**5.** $x \ge y^2 - 4$
$y \le 2x - 6$

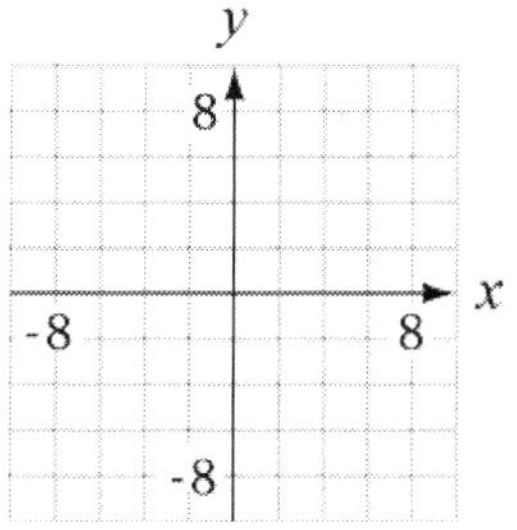


**6.** $\dfrac{x^2}{36} + \dfrac{y^2}{16} < 1$
$x^2 + y^2 > 4$

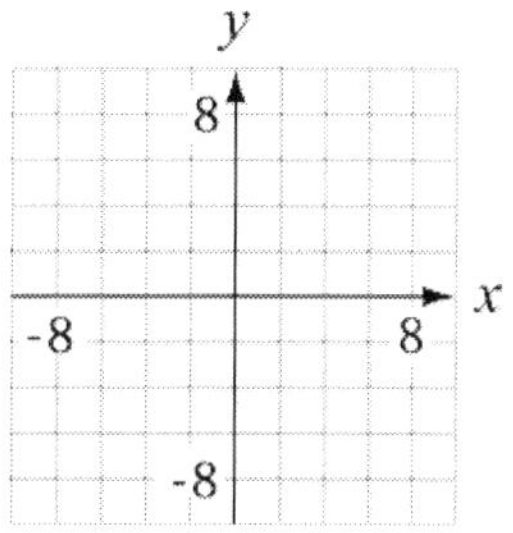

# CHAPTER 1

## Objective 1.1.1

| | | | | | | |
|---|---|---|---|---|---|---|
| **1.** > | **2.** < | **3.** > | **4.** < | **5.** > | **6.** < | **7.** False |
| **8.** False | **9.** True | **10.** False | **11.** False | **12.** False | **13.** False | **14.** False |
| **15.** True | **16.** {1, 2, 3} | **17.** {–3, –2, –1} | **18.** {5, 4, 3, 2, 1} | **19.** –6, –3 | **20.** –5, –1, 3 | |
| **21.** –20, 0 | **22.** 2, 3, 7, 18 | **23.** –26, –19, –7, –5, 2, 6 | | **24.** 0, 3, 6 | | |

## Objective 1.1.2

| | | | | | | |
|---|---|---|---|---|---|---|
| **1.** –7 | **2.** 4 | **3.** 5 | **4.** –34 | **5.** 28 | **6.** –66 | **7.** –3 |
| **8.** –7 | **9.** 5 | **10.** 13 | **11.** –4 | **12.** 15 | **13.** 16 | **14.** –19 |
| **15.** –19 | **16.** < | **17.** > | **18.** > | **19.** < | **20.** < | **21.** < |
| **22.** > | **23.** < | **24.** > | | | | |

## Objective 1.2.1

| | | | | | | |
|---|---|---|---|---|---|---|
| **1.** –2 | **2.** –2 | **3.** –10 | **4.** 7 | **5.** 2 | **6.** –18 | **7.** –7 |
| **8.** –14 | **9.** –5 | **10.** –6 | **11.** 4 | **12.** –4 | **13.** –5 | **14.** –25 |
| **15.** –5 | **16.** 10 | **17.** 36 | **18.** –4 | **19.** –92 | **20.** –39 | **21.** –9 |
| **22.** –1 | **23.** –17 | **24.** –27 | **25.** –3 | **26.** –23 | **27.** –9 | |

## Objective 1.2.2

| | | | | | | |
|---|---|---|---|---|---|---|
| **1.** 6 | **2.** 10 | **3.** –10 | **4.** –9 | **5.** –12 | **6.** 10 | **7.** 4 |
| **8.** –8 | **9.** 4 | **10.** –28 | **11.** –8 | **12.** –20 | **13.** –7 | **14.** 12 |
| **15.** –7 | **16.** –8 | **17.** 8 | **18.** 1 | **19.** –9 | **20.** 54 | **21.** –129 |
| **22.** 14 | **23.** –5 | **24.** –24 | **25.** 64 | **26.** –83 | **27.** 6 | |

## Objective 1.2.3

| | | | | | | |
|---|---|---|---|---|---|---|
| **1.** 44 | **2.** –8 | **3.** –12 | **4.** 14 | **5.** –16 | **6.** –24 | **7.** –42 |
| **8.** –20 | **9.** 28 | **10.** 18 | **11.** 0 | **12.** –63 | **13.** –90 | **14.** –57 |
| **15.** 52 | **16.** –108 | **17.** –162 | **18.** 192 | **19.** –180 | **20.** –42 | **21.** –120 |
| **22.** –108 | **23.** 48 | **24.** –210 | **25.** 56 | **26.** –180 | **27.** –48 | |

## Objective 1.2.4

| | | | | | | |
|---|---|---|---|---|---|---|
| **1.** –3 | **2.** –6 | **3.** 8 | **4.** –1 | **5.** –7 | **6.** –5 | **7.** –6 |
| **8.** –8 | **9.** 9 | **10.** 6 | **11.** 7 | **12.** –10 | **13.** 5 | **14.** 12 |
| **15.** 12 | **16.** 15 | **17.** 15 | **18.** –17 | **19.** –8 | **20.** –16 | **21.** –16 |
| **22.** –7 | **23.** 14 | **24.** –41 | **25.** 15 | **26.** 9 | **27.** –14 | |

## Objective 1.2.5

| | | | | | | |
|---|---|---|---|---|---|---|
| **1.** 9250 m | **2.** 107.2°F | **3.** 107°C | **4.** 379°C | **5.** 7°F | **6.** 5°C | **7.** –3°C |
| **8.** –1°C | **9.** 14 shots | **10.** 20 shots | | | | |

## Objective 1.3.1

| | | | | | | |
|---|---|---|---|---|---|---|
| **1.** 0.375 | **2.** $0.\overline{7}$ | **3.** $0.08\overline{3}$ | **4.** $0.4\overline{6}$ | **5.** 0.3125 | **6.** 0.15 | **7.** 0.45 |
| **8.** 0.95 | **9.** 0.12 | **10.** 0.28 | **11.** $0.3\overline{6}$ | **12.** 0.5 | | |

## Objective 1.3.2

| | | | | | | |
|---|---|---|---|---|---|---|
| **1.** $-\frac{1}{5}$ | **2.** $\frac{3}{8}$ | **3.** $\frac{1}{20}$ | **4.** $-\frac{9}{50}$ | **5.** $\frac{36}{77}$ | **6.** $-\frac{4}{15}$ | **7.** 4.26 |
| **8.** –3.24 | **9.** 2.136 | **10.** $1\frac{2}{3}$ | **11.** $-1\frac{1}{2}$ | **12.** $-\frac{5}{6}$ | **13.** $-2\frac{1}{12}$ | **14.** $1\frac{1}{8}$ |
| **15.** $-\frac{3}{4}$ | **16.** $-\frac{3}{5}$ | **17.** $1\frac{1}{3}$ | **18.** $1\frac{1}{3}$ | **19.** $-\frac{1}{12}$ | **20.** –4.5 | **21.** $\frac{1}{2}$ |
| **22.** –0.163 | **23.** 0.014 | **24.** $-\frac{23}{12}$ | | | | |

## Objective 1.3.3

| | | | | | | |
|---|---|---|---|---|---|---|
| **1.** $-\frac{1}{20}$ | **2.** $-\frac{1}{20}$ | **3.** $\frac{1}{24}$ | **4.** $-\frac{16}{45}$ | **5.** $-\frac{9}{16}$ | **6.** $\frac{1}{24}$ | **7.** $\frac{7}{9}$ |
| **8.** –3.337 | **9.** 9.086 | **10.** $\frac{2}{3}$ | **11.** $-\frac{35}{24}$ | **12.** –57.49 | **13.** $\frac{7}{12}$ | **14.** $\frac{4}{9}$ |
| **15.** $-\frac{29}{48}$ | **16.** $-1\frac{1}{9}$ | **17.** $\frac{11}{24}$ | **18.** $\frac{29}{30}$ | **19.** $-\frac{11}{36}$ | **20.** $\frac{11}{24}$ | **21.** –4.88 |
| **22.** 0 | **23.** $-\frac{1}{8}$ | **24.** $\frac{5}{4}$ | | | | |

## Objective 1.3.4

| | | | | | | |
|---|---|---|---|---|---|---|
| **1.** $\frac{3}{10}$; 0.30 | **2.** $\frac{11}{25}$; 0.44 | **3.** $1\frac{1}{10}$; 1.10 | **4.** $1\frac{2}{5}$; 1.40 | **5.** $\frac{21}{100}$; 0.21 | **6.** $\frac{3}{20}$; 0.15 | **7.** $\frac{2}{9}$ |
| **8.** $\frac{11}{250}$ | **9.** $\frac{137}{400}$ | **10.** 0.038 | **11.** 0.1617 | **12.** 0.3207 | **13.** 118% | **14.** 205.4% |
| **15.** 0.9% | **16.** 60% | **17.** 133.3% | **18.** 375% | **19.** 52% | **20.** $21\frac{3}{7}$% | **21.** 312.5% |

## Objective 1.4.1

| | | | | | | |
|---|---|---|---|---|---|---|
| **1.** 16 | **2.** 125 | **3.** –25 | **4.** –81 | **5.** 16 | **6.** –27 | **7.** 25 |
| **8.** 16 | **9.** –8 | **10.** –27 | **11.** 1728 | **12.** 256 | **13.** –4 | **14.** 8 |
| **15.** 400 | **16.** –320 | **17.** 576 | **18.** –48 | **19.** –12 | **20.** 24 | **21.** –180 |
| **22.** 216 | **23.** –432 | **24.** –72 | | | | |

## Objective 1.4.2

| | | | | | | |
|---|---|---|---|---|---|---|
| **1.** 3 | **2.** 14 | **3.** 8 | **4.** 11 | **5.** 20 | **6.** 19 | **7.** 5 |
| **8.** 5 | **9.** –16 | **10.** –5 | **11.** –13 | **12.** 0 | **13.** $-\frac{3}{2}$ | **14.** 6 |
| **15.** 10 | **16.** –2 | **17.** 6 | **18.** 1 | **19.** –49 | **20.** 3 | **21.** 8 |
| **22.** 9 | **23.** 8 | **24.** 1.305 | | | | |

## Objective 1.5.1

| | | | | | | |
|---|---|---|---|---|---|---|
| **1.** 13 | **2.** 2 | **3.** 25 | **4.** 8 | **5.** 3 | **6.** 22 | **7.** 49 |
| **8.** 7 | **9.** 2 | **10.** 25 | **11.** –10 | **12.** 1 | **13.** –32 | **14.** 17 |
| **15.** 15 | **16.** $-\frac{1}{2}$ | **17.** –7 | **18.** 1 | **19.** 22 | **20.** –3 | **21.** –8 |
| **22.** 4 | **23.** 58 | **24.** 1 | | | | |

## Objective 1.5.2

**1.** 6 **2.** 8 **3.** 1 **4.** $-x$ **5.** 0 **6.** $wz$ **7.** 20
**8.** 1 **9.** 0 **10.** 3 **11.** The Inverse Property of Addition **12.** The Distributive Property
**13.** The Multiplication Property of Zero **14.** The Multiplication Property of One
**15.** The Associative Property of Addition **16.** The Addition Property of Zero
**17.** The Associative Property of Multiplication **18.** The Inverse Property of Multiplication
**19.** The Commutative Property of Addition **20.** The Commutative Property of Multiplication

## Objective 1.5.3

**1.** $11x$ **2.** $21x$ **3.** $5a$ **4.** $7a$ **5.** $2ab$ **6.** $5xy$ **7.** $-2xy$
**8.** $-4xy$ **9.** $5ab$ **10.** 0 **11.** $-\frac{7}{12}x$ **12.** $-\frac{2}{5}x$ **13.** $-\frac{1}{3}x^2$ **14.** $\frac{1}{8}x$
**15.** $-\frac{17}{12}x^2$ **16.** $8a$ **17.** $-2a$ **18.** $-y^2$ **19.** $x+2y$ **20.** $7y-5x$ **21.** $7x$
**22.** $6a-12b$ **23.** $14a-13b$ **24.** $7y-11x$ **25.** $-2x^2-x$

## Objective 1.5.4

**1.** $10x$ **2.** $24x$ **3.** $-6a$ **4.** $12y$ **5.** $56y$ **6.** $-12x^2$ **7.** $-63x^2$
**8.** $x^2$ **9.** $x^2$ **10.** $a$ **11.** $x$ **12.** $x$ **13.** $a$ **14.** $a$
**15.** $x$ **16.** $x$ **17.** $y$ **18.** $n$ **19.** $3x$ **20.** $4x$ **21.** $-4x$
**22.** $-3x$ **23.** $-28a^2$ **24.** $-15x^2$ **25.** $5y$ **26.** $8y$ **27.** $4y$

## Objective 1.5.5

**1.** $-x-5$ **2.** $6x-3$ **3.** $8x-12$ **4.** $-3a-6$ **5.** $-4a-48$ **6.** $-4y+10$ **7.** $-12y+16$
**8.** $-6x^2-36$ **9.** $-20x^2-8$ **10.** $-10y^2+15$ **11.** $-21y^2+70$ **12.** $-6x^2-6y^2$ **13.** $-5x^2+5y^2$ **14.** $-3x^2+6y^2$
**15.** $-10a^2+16b^2$ **16.** $6x^2+6x-42$ **17.** $3x^2-6x-15$ **18.** $-2y^2-6y+10$
**19.** $-3a^2-3a-12$ **20.** $14x^2-28x-42$ **21.** $6x^2-8x+2$ **22.** $2x^2+4xy-8y^2$
**23.** $6x^2-9xy-12y^2$ **24.** $-2a^2-4a+3$ **25.** $9b^2+5b-8$

## Objective 1.5.6

**1.** $-x-21$ **2.** $-a-1$ **3.** $13-10x$ **4.** $17-6x$ **5.** $-2-5y$ **6.** $7n-6$ **7.** $4x-10$
**8.** $-2x+26$ **9.** $x-17$ **10.** $4y+2$ **11.** $12y-33$ **12.** $-x+5y$ **13.** $2a+5b$ **14.** $-6x+18$
**15.** $20x+90$ **16.** $3x-18$ **17.** $4x-48$ **18.** $-4x+16$ **19.** $10x+5$ **20.** $-5x+30$ **21.** $22x-48$
**22.** $-4x+49$ **23.** $a+2b$ **24.** $8b$ **25.** $13x-24y$ **26.** $22x-y$

## Objective 1.6.1

**1.** $6+x$ **2.** $y-10$ **3.** $12-x$ **4.** $t+6$ **5.** $a-1$ **6.** $y-5$ **7.** $\frac{z}{8}$
**8.** $11m$ **9.** $x^2+10$ **10.** $15-x^3$ **11.** $5(n+8)$ **12.** $\frac{2}{3}x+6$ **13.** $x+3x$ **14.** $\frac{-6}{y}$
**15.** $-2t$ **16.** $2(x+8)$ **17.** $5(y+6)$ **18.** $\frac{10}{y+5}$ **19.** $\frac{1}{4}y^2+12$ **20.** $-1m+16$ **21.** $x+\frac{x}{2}$
**22.** $\frac{3}{5}(10w)$ **23.** $y-2y$ **24.** $7(x+3)$

## Objective 1.6.2

**1.** $4n-n$; $3n$ **2.** $5n+n$; $6n$ **3.** $(n+12)-3$; $n+9$ **4.** $n+(11-n)$; $11$ **5.** $n+(n+7)$; $2n+7$

**6.** $n-(n+5)$; $-5$ **7.** $\frac{2}{5}n+\frac{3}{10}n$; $\frac{7}{10}n$ **8.** $\frac{3}{8}n-\frac{1}{4}n$; $\frac{1}{8}n$ **9.** $(n+9)-3$; $n+6$

**10.** $(n+3)+4$; $n+7$ **11.** $4(5n+6)$; $20n+24$ **12.** $11n-7n$; $4n$ **13.** $(n+10)+12$; $n+22$

**14.** $15-(n+5)$; $10-n$ **15.** $3(n+n+1)$; $6n+3$ **16.** $(n+n+1)+9$; $2n+10$ **17.** $\frac{1}{2}(n+n+2)$; $n+1$

**18.** $(n-9)+(n+20)$; $2n+11$ **19.** $5n-n$; $4n$ **20.** $6(n^2+4)$; $6n^2+24$

## Objective 1.6.3

**1.** gallons of paint in first container: $g$; gallons of paint in the second container: $10-g$
**2.** length of first piece: $L$; length of second piece: $35-L$ **3.** speed of corporate jet: $s$; speed of commercial jet: $3s$
**4.** amount at 3.5%: $A$; amount at 4%: $4200-A$ **5.** height of triangle: $h$; base of triangle: $h+4$
**6.** width: $w$; length: $4w$ **7.** length of shorter piece: $L$; length of longer piece: $8-L$
**8.** width of rectangle: $w$; length of rectangle: $2w-3$ **9.** number of apple trees: $N$; number of cherry trees: $\frac{1}{5}N$
**10.** number of full-time faculty: $F$; number of adjunct faculty: $F+206$

# CHAPTER 2

## Objective 2.1.1

**1.** Yes **2.** Yes **3.** No **4.** Yes **5.** Yes **6.** Yes **7.** No
**8.** No **9.** Yes **10.** No **11.** Yes **12.** Yes **13.** Yes **14.** Yes
**15.** No **16.** Yes **17.** Yes **18.** No **19.** Yes **20.** Yes **21.** No
**22.** Yes **23.** Yes **24.** No

## Objective 2.1.2

**1.** 5 **2.** 3 **3.** 15 **4.** 14 **5.** 9 **6.** –5 **7.** –8
**8.** 2 **9.** 3 **10.** 4 **11.** 2 **12.** 0 **13.** –5 **14.** 2
**15.** –6 **16.** 2 **17.** 6 **18.** –8 **19.** 10 **20.** 15 **21.** –6
**22.** 16 **23.** –1 **24.** $\frac{1}{4}$ **25.** $\frac{3}{5}$ **26.** $-\frac{5}{6}$ **27.** $-\frac{7}{6}$

## Objective 2.1.3

**1.** 3 **2.** 6 **3.** –5 **4.** –5 **5.** –3 **6.** 12 **7.** 7
**8.** 6 **9.** –7 **10.** 3 **11.** –6 **12.** –4 **13.** 12 **14.** 20
**15.** –12 **16.** –15 **17.** 14 **18.** –20 **19.** –16 **20.** 8 **21.** –6
**22.** –30 **23.** 12 **24.** 16 **25.** $\frac{3}{2}$ **26.** 2 **27.** 5

## Objective 2.1.4

**1.** 16% **2.** 20% **3.** 4.8 **4.** 27 **5.** 400 **6.** 32 **7.** 6
**8.** 40% **9.** 40 **10.** 64 **11.** 500% **12.** 400% **13.** 7.2 **14.** 3.78
**15.** 18.48 **16.** 20% **17.** 52 cars **18.** 140 employees **19.** 600 computers **20.** 20%
**21.** 57,600 building permits

## Objective 2.2.1

**1.** 6 **2.** 7 **3.** 4 **4.** 7 **5.** –3 **6.** –5 **7.** –7
**8.** 2 **9.** 2 **10.** 1 **11.** 5 **12.** 6 **13.** 2 **14.** 2
**15.** 7 **16.** –4 **17.** –7 **18.** 4 **19.** 9 **20.** 2 **21.** 2
**22.** 3 **23.** 1 **24.** 2 **25.** 6 **26.** 4 **27.** –9

## Objective 2.2.2

**1.** 7 **2.** 4 **3.** 8 **4.** 5 **5.** 4 **6.** –2 **7.** –3
**8.** –2 **9.** –3 **10.** 2 **11.** –3 **12.** –2 **13.** –2 **14.** –4
**15.** –6 **16.** 0 **17.** 0 **18.** –1 **19.** –2 **20.** 2 **21.** –1
**22.** –3 **23.** 6 **24.** 10 **25.** –2 **26.** –3 **27.** 6

## Objective 2.2.3

**1.** 2 **2.** 3 **3.** –6 **4.** 7 **5.** –1 **6.** 1 **7.** –2
**8.** 3 **9.** 2 **10.** –2 **11.** –7 **12.** 6 **13.** 8 **14.** –3
**15.** 0 **16.** –1 **17.** –12 **18.** 2 **19.** –3 **20.** 4 **21.** –3
**22.** 2 **23.** 2 **24.** 9 **25.** –4 **26.** –2 **27.** 0

## Objective 2.2.4

**1.** $n + 9 = 15; n = 6$ **2.** $n + 2 = 11; n = 9$ **3.** $n - 6 = 10; n = 16$ **4.** $\frac{1}{2}n = 9; n = 18$

**5.** $4n + 5 = 13; n = 2$ **6.** $2(n + 3) = 12; n = 3$ **7.** $3n - 8 = 7; n = 5$ **8.** $\frac{n}{6} = -2; n = -12$

**9.** $7n + 3 = -18; n = -3$ **10.** $5n + n = 24; n = 4$ **11.** $6n - 3n = 12; n = 4$ **12.** $30 = 5 - 5n; n = -5$
**13.** $9 = 2n + 1; n = 4$ **14.** $2n = 4(21 - n)$; 7 and 14 **15.** $5n = 2n + 18; n = 6$ **16.** $4(n + 3) = 5n - 2; n = 14$

## Objective 2.3.1

**1.** 12 lb at \$4.50/lb and 8 lb at \$3/lb **2.** 12 lb **3.** 18 kg at \$3/kg and 12 kg at \$4/kg **4.** 60 oz
**5.** 30 lb **6.** \$3.20 **7.** 15 kg of peanuts and 30 kg of cashews **8.** \$0.94
**9.** 2250 bushels of soybeans **10.** \$3.24

## Objective 2.3.2

**1.** 180 mph; 210 mph **2.** 40 mph; 60 mph **3.** 220 mi **4.** 54 mph; 42 mph **5.** $3\frac{3}{4}$ h

**6.** 960 mi **7.** 4 h at 55 mph; 2 h at 40 mph **8.** 60 mi **9.** $1\frac{1}{2}$ h **10.** 345 mph; 375 mph

## Objective 2.4.1

**1.** length is 54 m; width is 18 m **2.** 10 ft, 13, ft, 19 ft **3.** length is 27 m; width is 11 m **4.** 40 ft, 80 ft, 60 ft
**5.** 60 cm, 45 cm, 60 cm **6.** 60 cm, 30 cm, and 90 cm **7.** length is 33 m; width is 26 m
**8.** length is 40 ft; width is 24 ft **9.** 20 ft, 20 ft, 13 ft **10.** 4 m

## Objective 2.4.2

**1.** 69° **2.** 0° **3.** 81° **4.** 101° **5.** 12° **6.** 14° **7.** 47°
**8.** 5 **9.** $a = 123°, b = 57°$ **10.** $a = 39°, b = 141°$ **11.** 20° **12.** 25°

## Objective 4.2.3

**1.** 50° **2.** 37° **3.** 55° **4.** 69° **5.** 14° **6.** 29° **7.** 30°

**8.** 57°

### Objective 2.5.1

**1.** \$6.75 **2.** \$25 **3.** \$10 **4.** 40% **5.** \$4.26 **6.** \$32.40 **7.** 50%
**8.** 50% **9.** $33\frac{1}{3}\%$ **10.** 45% **11.** 80% **12.** \$144

### Objective 2.5.2

**1.** \$160 **2.** \$24 **3.** \$40 **4.** \$70 **5.** 20% **6.** 25% **7.** \$40
**8.** \$150 **9.** 15% **10.** \$17.40 **11.** \$36 **12.** $22\frac{1}{2}\%$

### Objective 2.6.1

**1.** \$3000 at 10%; \$5000 at 12% **2.** \$5000 at 9%; \$9000 at 15% **3.** \$6000 **4.** \$4800 at 18%; \$7200 at 12%
**5.** \$4000 at 15%; \$3000 at 20% **6.** \$6000 **7.** \$2,000,000 **8.** \$30,000

### Objective 2.6.2

**1.** 150 ml of 12% solution and 300 ml of 9% solution **2.** 25 g **3.** 25 oz
**4.** 90 L of pure maple syrup and 30 L of 80% maple syrup **5.** 42 lb of 15% fat and 18 lb of 25% fat **6.** 30 lb
**7.** 66% **8.** 69.5% **9.** 200 lb of 25% polyester and 100 lb of 55% polyester **10.** 150 lb

### Objective 2.7.1

**1.** 3 **2.** 7 **3.** –5 **4.** –17 **5.** 87 **6.** –72 **7.** –66
**8.** –52 **9.** 120 **10.** 36 **11.** –3, 4 **12.** –8, –2 **13.** 60, –13, –28 **14.** 48, 5, 9
**5.** > **6.** > **7.** < **8.** < **9.** > **10.** < **11.** < **12.** > **13.** < **14.** > **15.** > **16.** <

### Objective 2.7.1

**1.** $\{-18, -12, -6\}$ **2.** $\{4, 8, 12, 16, 20\}$ **3.** $\{23, 29\}$ **4.** $\{x \mid x \geq -6\}$ **5.** $\{x \mid -1 \leq x \leq 1\}$
**6.** $\{x \mid -10 < x < 10, x \text{ is an integer}\}$ **7.** $A \cup B = \{-2, -1, 0, 1, 2\}$ **8.** $A \cup B = \{-3, -1, 0, 1, 3\}$
**9.** $A \cup B = \{-2, -1, 0, 2\}$ **10.** $A \cap B = \{0\}$ **11.** $A \cap B = \{5, 10\}$ **12.** $A \cap B = \{1, 4, 9\}$

**13.** 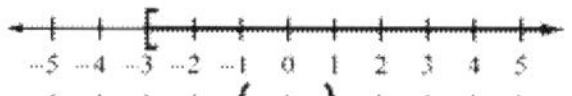
**14.** 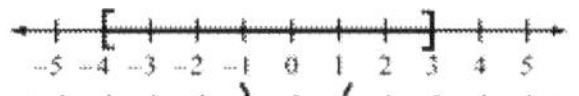
**15.** 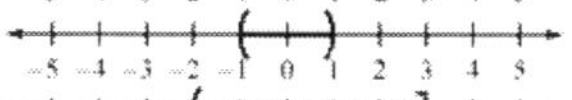
**16.** 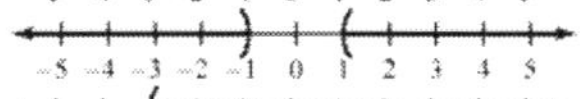
**17.** 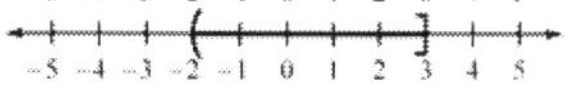
**18.** 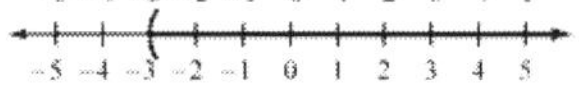

### Objective 2.7.2

**1.** \$6400 @ 4.5%; \$3600 @ 6% **2.** \$8000 @ 10%; \$6000 @ 14.5% **3.** \$2500 **4.** \$6000
**5.** \$6500 @ 11%; \$5500 @13% **6.** \$14,250 @ 15%; \$23,750 @9% **7.** \$14,000 **8.** \$10,000
**9.** \$6000 **10.** \$14,500

### Objective 2.7.3

**1.** 46% **2.** 28% **3.** 24 ml **4.** 50 oz **5.** 20 lb of 24%; 60 lb of 16%
**6.** 11.25 L of 70%; 18.75 L of 30% **7.** 75 g **8.** 640 lb **9.** 45 L of 70%; 75 L of 30%
**10.** 450 lb of 20%; 150 lb of 80% **11.** 30 ml **12.** 75 ml

## Objective 2.7.4

**1.** 12 **2.** -19 **3.** 169.6 miles **4.** Between -5°C and 20°C **5.** \$32,000 **6.** 50 checks
**7.** Between 53 and 98 **8.** Answers vary; any three consecutive odd integers beginning between 25 and 31

## Objective 2.8.1

**1.** $\{x \mid x < 6\}$ **2.** $\{x \mid x \geq -3\}$ **3.** $\{x \mid x \leq 5\}$ **4.** $\{x \mid x > 5\}$ **5.** $\{x \mid x < -3\}$ **6.** $\{x \mid x \geq 4\}$ **7.** $\{x \mid x > 2\}$
**8.** $\{x \mid x \geq -2\}$ **9.** $\{x \mid x > 3\}$ **10.** $\{x \mid x < 6\}$ **11.** $\{x \mid x \leq 3\}$ **12.** $\{x \mid x \leq -2\}$ **13.** $\{x \mid x > -3\}$ **14.** $\{x \mid x < -5\}$
**15.** $\{x \mid x > -9\}$ **16.** $\{x \mid x < -2\}$ **17.** $\{x \mid x \geq 4\}$ **18.** $\{x \mid x < 2\}$ **19.** $\{x \mid x \leq -4\}$ **20.** $\{x \mid x \geq -3\}$ **21.** $\{x \mid x < 1\}$
**22.** $\{x \mid x > -8\}$ **23.** $\{x \mid x < -4\}$ **24.** $\{x \mid x \geq -5\}$

## Objective 2.8.2

**1.** $\{x \mid -3 < x < 5\}$ **2.** $\{x \mid -5 < x \leq 8\}$ **3.** $\{x \mid x \geq 6 \text{ or } x < 3\}$ **4.** $\{x \mid x < 4 \text{ or } x > 9\}$
**5.** $\{x \mid -3 < x < 6\}$ **6.** $\{x \mid -3 < x < 3\}$ **7.** $\{x \mid x \geq 4\}$ **8.** $\{x \mid x < -5\}$ **9.** $\varnothing$ **10.** $\{x \mid -2 < x < 2\}$
**11.** $\{x \mid x < 1 \text{ or } x > 3\}$ **12.** $\{x \mid x < -3 \text{ or } x > 3\}$ **13.** $(-3, 1)$ **14.** $(-2, 4)$ **15.** $(4, 5)$ **16.** $(-3, -1)$
**17.** $(-\infty, -2] \cup (3, \infty)$ **18.** $(-\infty, -1) \cup (3, \infty)$ **19.** $(4, \infty)$ **20.** $(-\infty, 2)$ **21.** $[-4, 3]$ **22.** $[3, 5]$
**23.** $[-2, 1)$ **24.** $\varnothing$

## Objective 2.8.3

**1.** 3 ft **2.** –19 **3.** less than 2501 miles **4.** $-5° < C < 20°$ **5.** \$32,000 or more
**6.** less than 30 checks **7.** $53 \leq N \leq 98$ **8.** 25, 27, 29 or 27, 29, 31 or 29, 31, 33

# CHAPTER 3

## Objective 3.1.1

**1.**
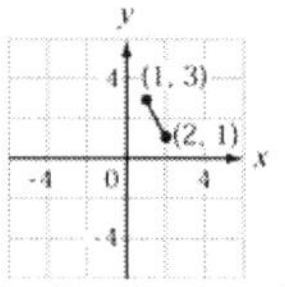


**2.**
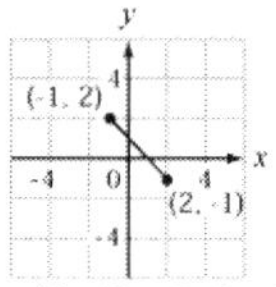


**3.**
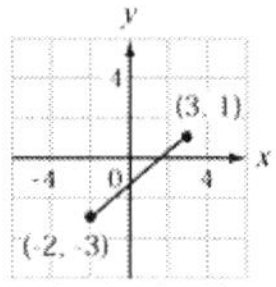


**4.**
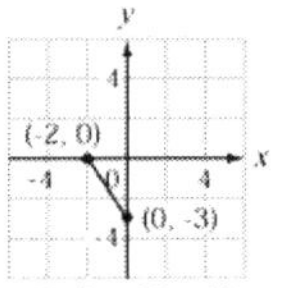


**5.** $A$ is (2, 0), $B$ is (–3, 1), $C$ is (1, 3), $D$ is (–2, -3) **6.** $A$ is (–4, 1), $B$ is (–3, –2), $C$ is (3, 2), $D$ is (1, 4)

**7.**
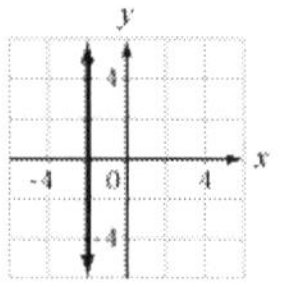

**8.**
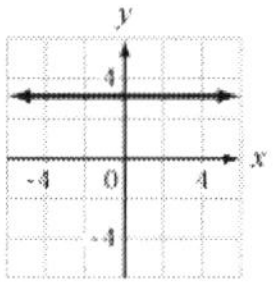

## Objective 3.1.2

**1.** 5.39 **2.** 5 **3.** 6.08 **4.** 4.47 **5.** $(-1, 0)$ **6.** $\left(\frac{5}{2}, \frac{3}{2}\right)$ **7.** $(0, 3)$

**8.** $\left(\frac{1}{2}, \frac{1}{2}\right)$ **9.** $(1, 1)$ **10.** $\left(\frac{3}{2}, -\frac{3}{2}\right)$

## Objective 3.1.3

1. 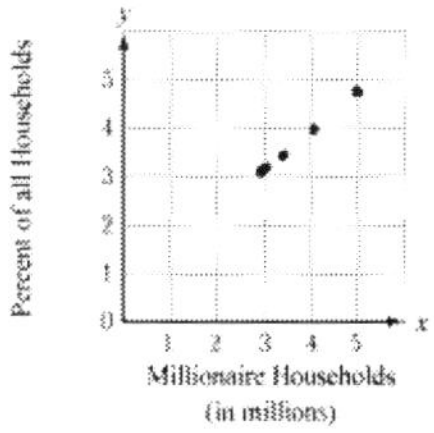


2. 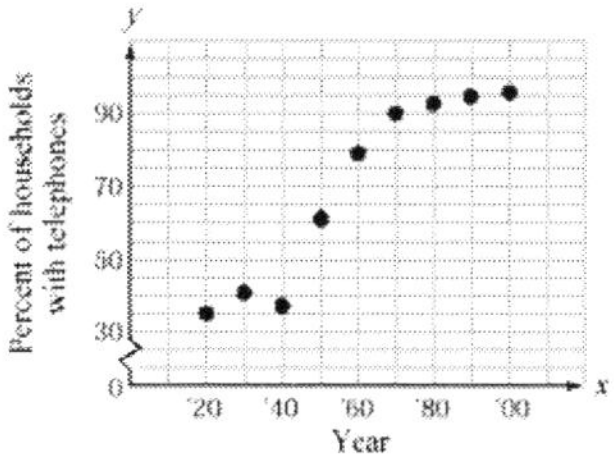


## Objective 3.1.4

**1.** 4° per hour **2.** $0.\overline{8}$ cents per year **3.** 2,435,000 acres per year **4.** Less than **5.** 3509 lobbyists per year
**6.** Less than

## Objective 3.2.1

**1.** Yes **2.** No **3.** –1 **4.** 5 **5.** 1 **6.** –1 **7.** 1
**8.** –1 **9.** 5 **10.** 1 **11.** $t^2+t-1$ **12.** Domain: {1, 2, 4, 5}; Range: {1, 3, 7}
**13.** Domain: {1, 3, 5, 6}; Range: {1, 2, 4} **14.** Domain: {–1, 1, 3}; Range: {–1, 7, 15}
**15.** Domain: {–1, 0, 3}; Range: $\left\{\frac{3}{5}, \frac{3}{4}, 1\right\}$ **16.** Domain: {–2, 1, 4}; Range: {11, –1, 5}
**17.** Domain: {–2, 0, 4}; Range: {–4, –1, 5}

## Objective 3.2.2

1. 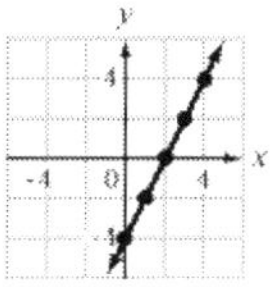
2. 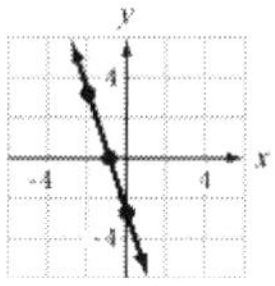
3. 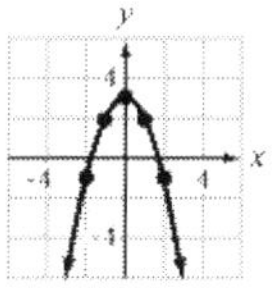
4. 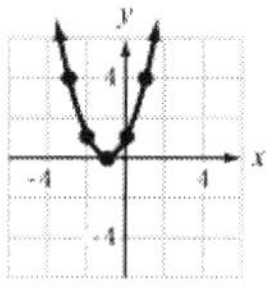
5. 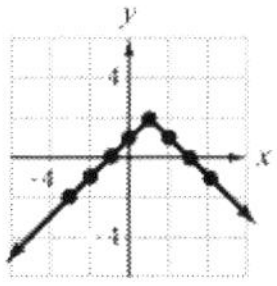
6. 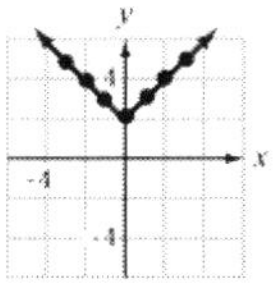

## Objective 3.2.3

**1.** Yes **2.** Yes **3.** No **4.** Yes **5.** No **6.** No **7.** Yes
**8.** Yes

## Objective 3.3.1

1. 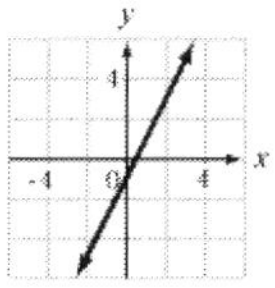
2. 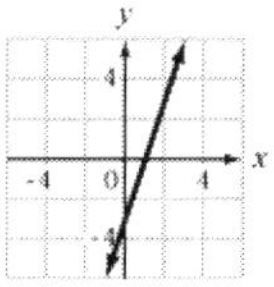
3. 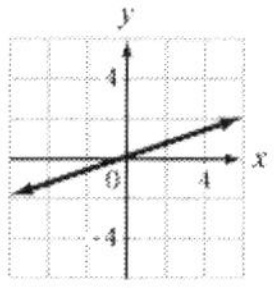
4. 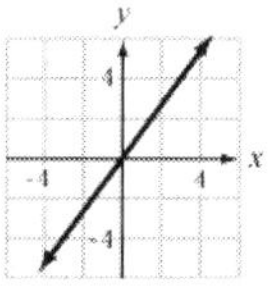
5. 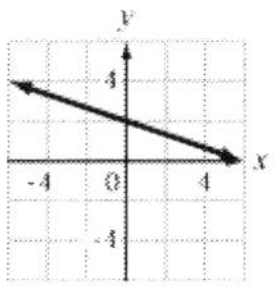
6. 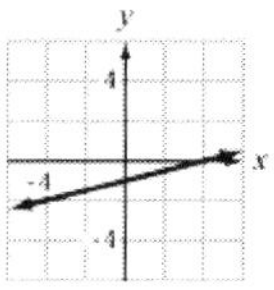
7. 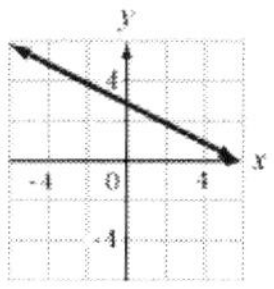
8. 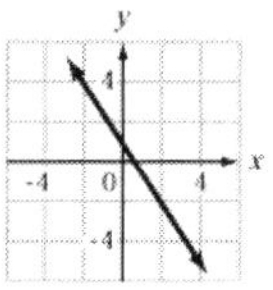

## Objective 3.3.2

**1.**

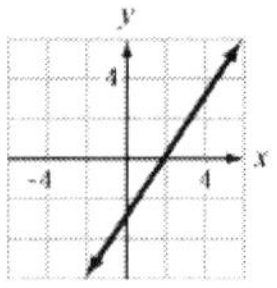


**2.**

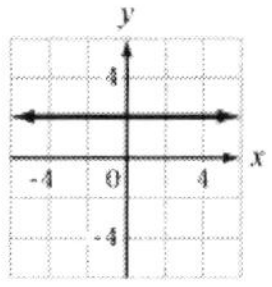


**3.** $x$-intercept: (5, 0)
$y$-intercept (0, –2)

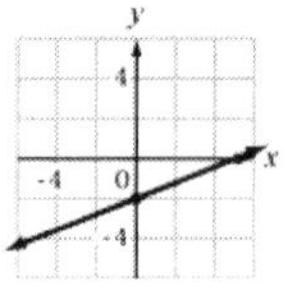


**4.** $x$-intercept: (2, 0)
$y$-intercept (0, –6)

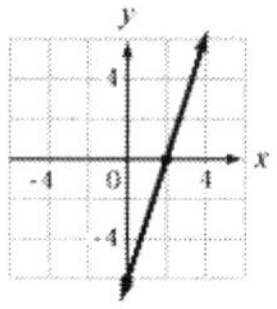


**5.** $x$-intercept: (0, 0)
$y$-intercept (0, 0)

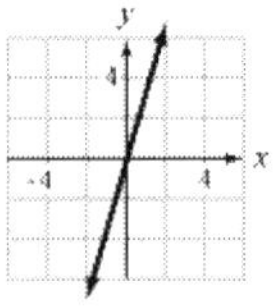


**6.** $x$-intercept: $\left(\frac{5}{2}, 0\right)$
$y$-intercept (0, –5)

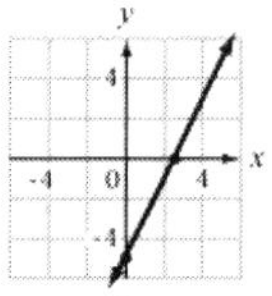


**7.** 3 **8.** $\frac{7}{2}$ **9.** $\frac{3}{2}$ **10.** $-\frac{25}{2}$

## Objective 3.3.3

**1.** Loren earns $160 for tutoring 16 hours.

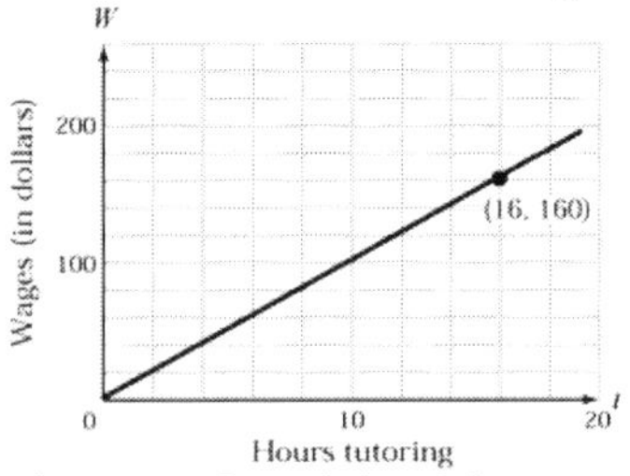


**2.** The roller coaster travels 594 ft in 6 s.

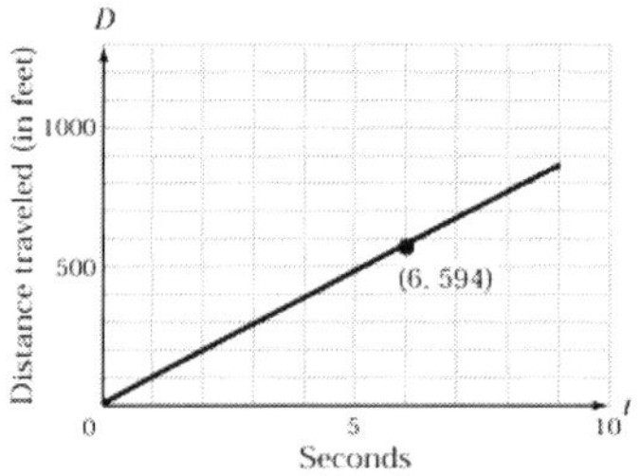


**3.** The cost of receiving 36 messages is $16.

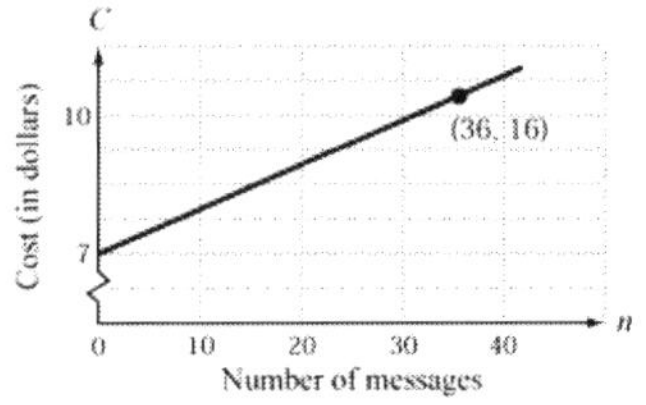


**4.** The cost of manufacturing 60 snow boards is $9000.

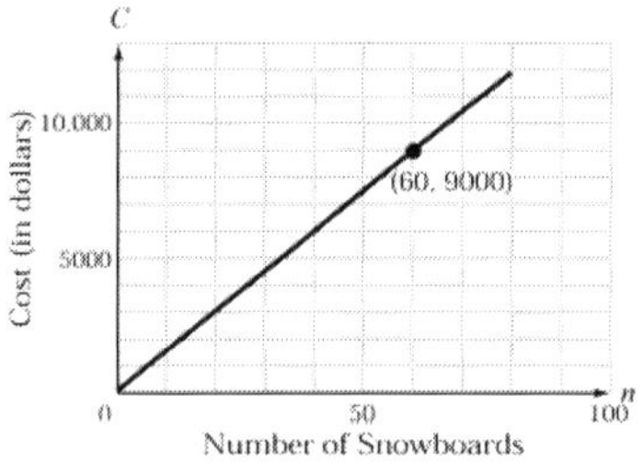


## Objective 3.4.1

**1.** $\frac{1}{2}$ **2.** $-\frac{1}{2}$ **3.** –3 **4.** 1 **5.** 1 **6.** $-\frac{1}{2}$ **7.** –1

**8.** $-\frac{1}{3}$ **9.** $-\frac{5}{3}$ **10.** $\frac{5}{2}$ **11.** 1 **12.** 2 **13.** –1 **14.** –5

**15.** –2 **16.** $\frac{1}{3}$ **17.** –3 **18.** –1 **19.** 0 **20.** $\frac{1}{4}$ **21.** 1

## Objective 3.4.2

1. 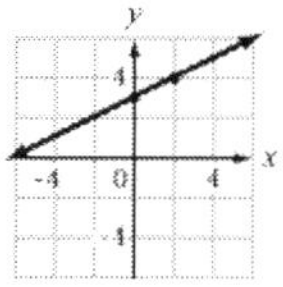
2. 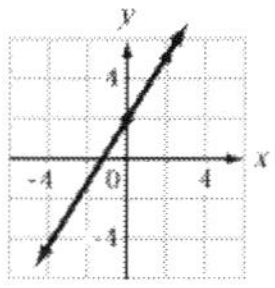
3. 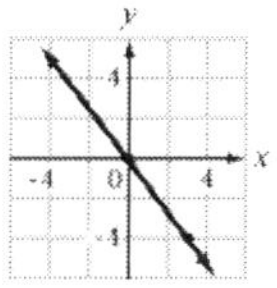
4. 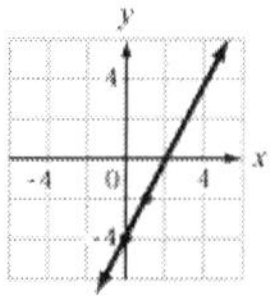
5. 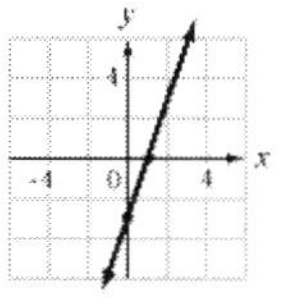
6. 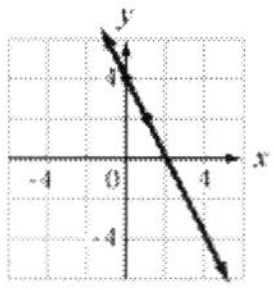
7. 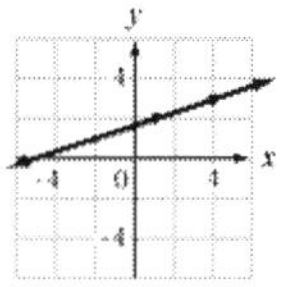
8. 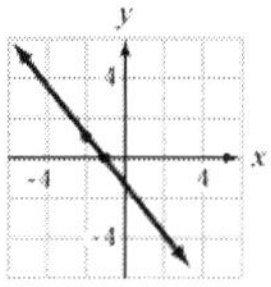

## Objective 3.5.1

**1.** $y = 2x + 4$ **2.** $y = 4x + 8$ **3.** $y = \frac{1}{3}x + \frac{5}{3}$ **4.** $y = -\frac{1}{3}x - 1$ **5.** $y = -x - 1$ **6.** $y = \frac{1}{2}x + \frac{7}{2}$ **7.** $y = \frac{2}{3}x + \frac{5}{3}$

**8.** $y = -\frac{1}{3}x + 3$ **9.** $y = -2x + 3$ **10.** $y = -\frac{2}{5}x + \frac{13}{5}$ **11.** $y = \frac{3}{4}x - \frac{11}{4}$ **12.** $y = \frac{1}{3}x - \frac{8}{3}$

**13.** $y = -\frac{1}{2}x - \frac{3}{2}$ **14.** $y = -\frac{1}{4}x$ **15.** $y = 2x - 8$ **16.** $y = -2x + 2$ **17.** $y = -\frac{1}{3}x + \frac{14}{3}$

**18.** $y = -\frac{3}{5}x + \frac{17}{5}$

## Objective 3.5.2

**1.** $y = -x + 5$ **2.** $y = -x + 4$ **3.** $y = \frac{2}{3}x + \frac{4}{3}$ **4.** $y = \frac{1}{2}x - \frac{1}{2}$ **5.** $y = \frac{1}{3}x + \frac{8}{3}$ **6.** $y = x$ **7.** $y = \frac{3}{2}x$

**8.** $y = \frac{4}{3}x + \frac{11}{3}$ **9.** $y = 2x - 2$ **10.** $y = 3x + 3$ **11.** $y = \frac{1}{2}x$ **12.** $y = -3x$ **13.** $y = -\frac{2}{3}x + \frac{2}{3}$

**14.** $y = -\frac{4}{3}x - \frac{4}{3}$ **15.** $y = -\frac{6}{5}x + \frac{8}{5}$ **16.** $y = -\frac{6}{7}x - \frac{11}{7}$ **17.** $y = -x + 2$ **18.** $y = -2$

## Objective 3.5.3

**1a.** $y = 95x + 35{,}000$ **1b.** \$177,500 **2a.** $y = -0.025x + 13$ **2b.** 8.5 gal **3a.** $y = -0.10x + 315{,}000$
**3b.** 82,500 cars **4a.** $y = 0.05x + 1200$ **4b.** \$4700 **5a.** $y = 22.5x$ **5b.** 112.5 Calories
**6a.** $y = 0.69x + 7.95$ **6b.** \$14.16 **7a.** $y = -3.5x + 100$ **7b.** 82.5°C **8.** $f(x) = -2x - 7$

## Objective 3.6.1

**1.** Yes **2.** No **3.** No **4.** Yes **5.** Yes **6.** Yes **7.** Yes

**8.** No **9.** Yes **10.** $y = -\frac{1}{3}x - \frac{17}{3}$ **11.** $y = \frac{1}{4}x - \frac{7}{4}$ **12.** $y = \frac{5}{3}x - \frac{11}{3}$

**13.** $y = -\frac{1}{2}x - \frac{1}{2}$ **14.** $y = -\frac{4}{3}x + \frac{8}{3}$ **15.** $y = \frac{2}{3}x - \frac{2}{3}$ **16.** $y = -3x + 2$

## Objective 3.7.1

1. 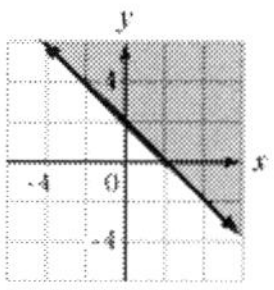
2. 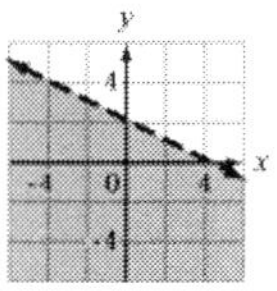
3. 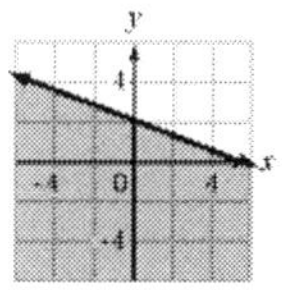
4. 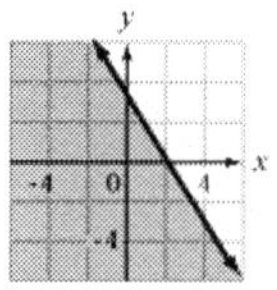

**5.** 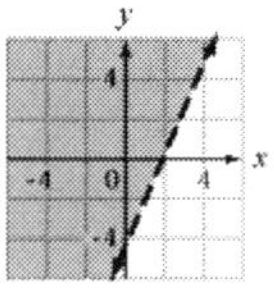


**6.** 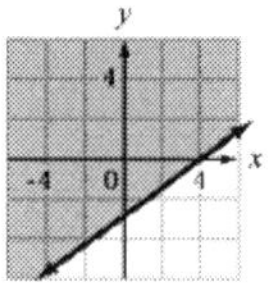


**7.** 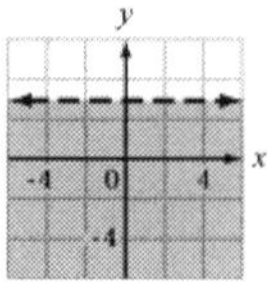


**8.** 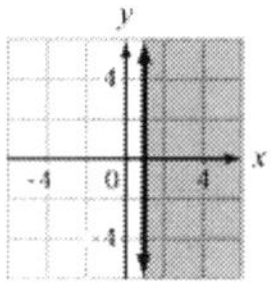


# CHAPTER 4

## Objective 4.1.1

**1.** The solution is (–3, 2).

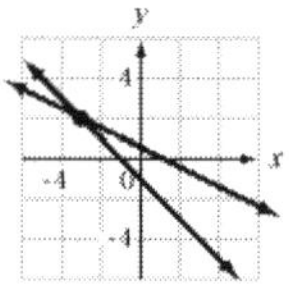


**2.** The solution is (2, –1).

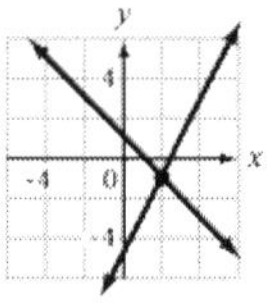


**3.** The solution is (–2, –3).

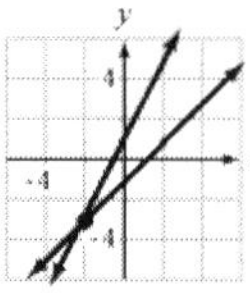


**4.** The solution is (3, 1).

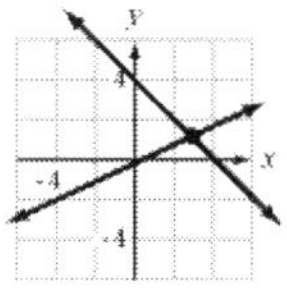


**5.** The solution is (2, –2).

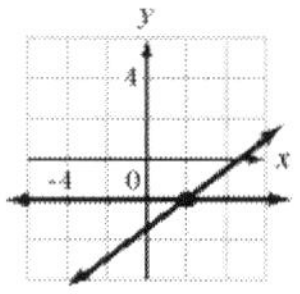


**6.** The solution is (3, 2).

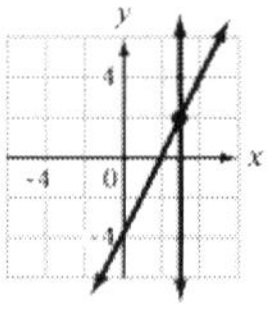


**7.** The solution is (2, 1).

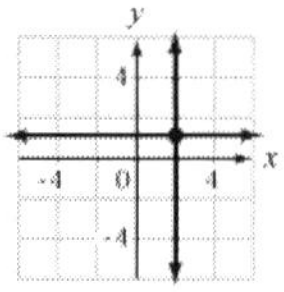


**8.** The solution is (–1, 3).

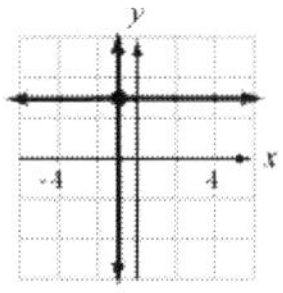


## Objective 4.1.2

**1.** (3, 2) **2.** (4, –1) **3.** (3, 1) **4.** (–2, 0) **5.** (5, 3) **6.** (2, 1) **7.** (1, –2)
**8.** (–2, 3) **9.** (2, 1) **10.** (–2, –1) **11.** (–1, 3) **12.** (2, –3) **13.** (–1, 4) **14.** (–2, –1)
**15.** (4, 1) **16.** $\left(1, -\frac{1}{2}\right)$ **17.** (2, –2) **18.** (1, 1) **19.** (–5, -2) **20.** (–4, –2) **21.** (1, 3)
**22.** (1, 1) **23.** (2, 5) **24.** (–2, 7)

## Objective 4.2.1

**1.** (4, 1) **2.** (3, 2) **3.** (4, –3) **4.** (1, –1) **5.** (–1, 3) **6.** (–2, 2) **7.** (5, 1)
**8.** (3, 5) **9.** (4, –2) **10.** (2, 5) **11.** (2, –1) **12.** (3, 2) **13.** $\left(\frac{1}{2}, 1\right)$ **14.** (–2, 1)
**15.** (1, –3) **16.** (5, –1) **17.** (–1, –2) **18.** (2, –3) **19.** (–1, 2) **20.** (4, 2) **21.** ( 2, 1)
**22.** (4, –1) **23.** $\left(-\frac{1}{7}, -\frac{25}{7}\right)$ **24.** (20, 15)

## Objective 4.2.2

**1.** (–2, 1, 3) **2.** (4, –2, 1) **3.** (–5, 3, –2) **4.** (–3, 4, 1) **5.** (3, –4, 2) **6.** (–1, 3, 4) **7.** (1, –2, –1)
**8.** (–2, 3, –1) **9.** $\left(\frac{162}{11}, \frac{65}{11}, -6\right)$ **10.** (1, –3, –4)

## Objective 4.3.1

**1.** 10 **2.** –7 **3.** 12 **4.** –11 **5.** –19 **6.** –13 **7.** 34
**8.** 1 **9.** 2 **10.** –50 **11.** –21 **12.** –32 **13.** –26 **14.** 32
**15.** 16 **16.** 7 **17.** –51 **18.** 2

### Objective 4.3.2

**1.** (1, 8) **2.** (0, –1) **3.** (4, 1) **4.** (3, –1) **5.** (3, 7) **6.** (1, –1) **7.** (2, 2)
**8.** (–3, –5) **9.** $\left(-\frac{1}{3}, 2\right)$ **10.** (0, 2) **11.** (–1, 1, 4) **12.** (1, 2, 5) **13.** (2, –1, 1) **14.** (–1, 1, 3)
**15.** (–1, 2, 1) **16.** (3, 1, 0)

### Objective 4.3.3

**1.** $\left[\begin{array}{cc|c} 1 & -3 & 6 \\ 0 & 1 & -3 \end{array}\right]$ **2.** $\left[\begin{array}{cc|c} 1 & 0 & 7 \\ 0 & 1 & 47 \end{array}\right]$ **3.** $\left(-2, \frac{2}{3}\right)$ **4.** $\left(-\frac{7}{2}, -3\right)$ **5.** (–3, 4) **6.** $\left(3, \frac{5}{2}\right)$
**7.** (–5, 3, 2) **8.** inconsistent **9.** (–1, –3, –8) **10.** (3, –2, 1)

### Objective 4.4.1

**1.** plane: 170 mph; wind: 35 mph **2.** plane: 520 mph; wind: 30 mph **3.** cruiser: 15 mph; current: 5 mph
**4.** motorboat: 15 mph; current: 3 mph **5.** plane: 180 mph; wind: 20 mph **6.** jet: 225 mph; wind: 25 mph
**7.** boat: 24 km/h; current: 4 km/h **8.** team: 12 mph; current: 3 mph **9.** plane: 140 mph; wind: 20 mph
**10.** plane: 315 mph; wind: 45 mph

### Objective 4.4.2

**1.** pine: \$0.20/ft; redwood: \$0.30/ft **2.** cinnamon: \$2/lb; spice: \$3.50/lb **3.** 10¢/unit **4.** \$12/yd
**5.** 40 quarters **6.** 18 dimes **7.** \$3/lb **8.** tin: \$3.50/lb; zinc: \$9/lb
**9.** orange trees: \$8; grapefruit trees: \$6 **10.** bulbs: 50¢; lights: \$2

### Objective 4.5.1

**1.**
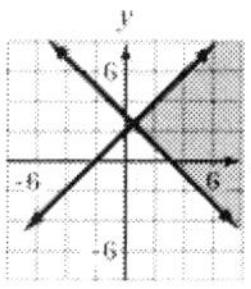
**2.**
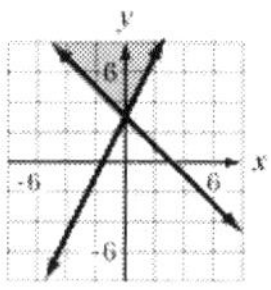
**3.**
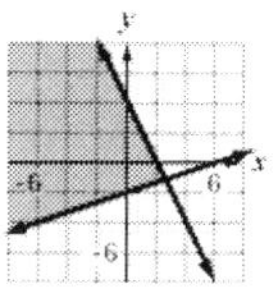
**4.**
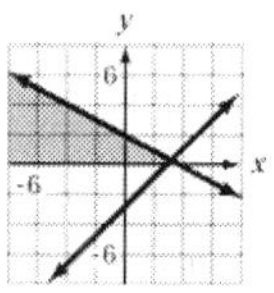
**5.**
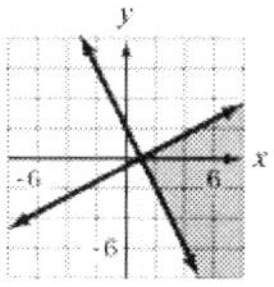
**6.**
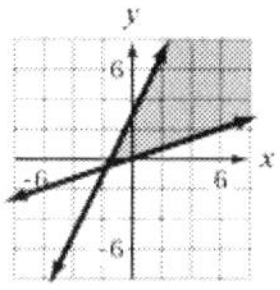
**7.**
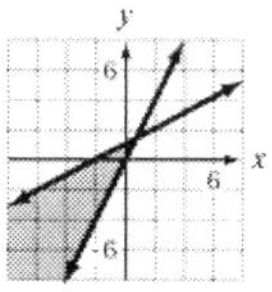
**8.**
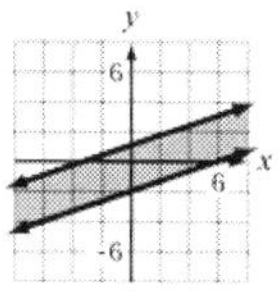

# CHAPTER 5

### Objective 5.1.1

**1.** –129 **2.** 22 **3a.** 6 **3b.** 7 **3c.** 4 **4.** not a polynomial **5a.** 1
**5b.** $-\sqrt{7}$ **5c.** 2 **6a.** $\pi$ **6b.** $\pi$ **6c.** 0
**7.**
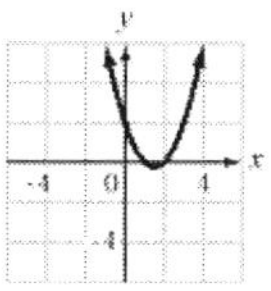
**8.**
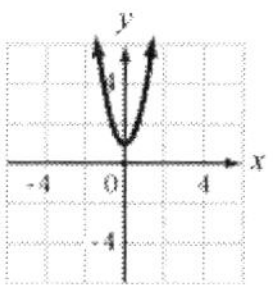
**9.**
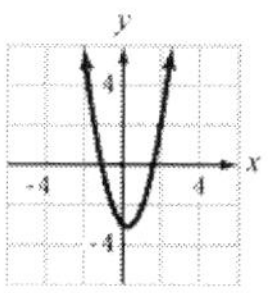

### Objective 5.1.2

**1.** $x^2 - 4x - 4$ **2.** $3x^3 + 2x^2 - 16x - 4$ **3.** $4x^2 - x - 4$ **4.** $-2x^2 + 3xy + y^2$ **5.** $x^{2n} + 5x^n + 3$

**6.** $5x^{2n}-2x^n+6n+1$ **7.** $4y^3+y^2-15y+1$ **8.** $3y^2-8y-11$ **9.** $6a^2-a+2$ **10.** $4x^4+2x^3-5x^2+3x$
**11.** $2x^4+4x^3-7x-5$ **12.** $-2a^{2n}-a^n-11$ **13.** $-2x^{2n}-x^n-11$ **14.** $-3x^3+9x^2+x-11$
**15.** $3a^3-8a^2b+8ab^2+11b^3$

## Objective 5.2.1

**1.** $3x^2$ **2.** $-2y^2$ **3.** $6x^2$ **4.** $16y^5$ **5.** $8a^5$ **6.** $-8a^{10}$ **7.** $x^3y^5$
**8.** $-12x^5y$ **9.** $-6a^6b^2$ **10.** $a^3b^5$ **11.** $-8x^3y^5$ **12.** $5a^3b^5$ **13.** $x^3y^5z$ **14.** $a^3b^3c^3$
**15.** $x^4y^4z$ **16.** $-a^4b^7$ **17.** $-x^4y^4$ **18.** $28y^7z^4$ **19.** $-6x^4y^2$ **20.** $a^4b^4c^6$ **21.** $x^5y^8z^3$
**22.** $24a^6b^6$ **23.** $-24x^5y^7$ **24.** $-6a^7b^6$ **25.** $8x^5y^8$ **26.** $-6a^{10}b^4$ **27.** $-30a^4b^3c^3$

## Objective 5.2.2

**1.** $x^4$ **2.** $x^6$ **3.** $y^8$ **4.** $x^{10}$ **5.** $y^{12}$ **6.** $x^6$ **7.** $8x^3$
**8.** $9y^2$ **9.** $-8x^3$ **10.** $9y^4$ **11.** $x^2y^4$ **12.** $x^8y^{12}$ **13.** $-8a^3b^6$ **14.** $x^{10}y^{10}$
**15.** $x^5y^{15}$ **16.** $x^{16}y^{12}$ **17.** $4x^4y^5$ **18.** $-a^6b^{12}$ **19.** $a^6b^6$ **20.** $-54y^7$ **21.** $a^4b^3$
**22.** $-12x^5y^4$ **23.** $-18x^4y^3$ **24.** $4a^3b^8$ **25.** $-54a^5b^6$ **26.** $-24a^6b^5$ **27.** $-243a^5b^8$

## Objective 5.3.1

**1.** $x^2+x$ **2.** $2y-y^2$ **3.** $-x^2-2x$ **4.** $2a^2-2a$ **5.** $3b^2+15b$ **6.** $-2x^3+2x^2$ **7.** $-4y^3-24y^2$
**8.** $-2x^4+3x^2$ **9.** $10x^3-4x^2$ **10.** $6y^2-3y^3$ **11.** $2x^2-3x$ **12.** $6x^2-3x$ **13.** $-x^3y+x^2y^3$
**14.** $-2x^2y^2+xy^3$ **15.** $x^3-2x^2+x$ **16.** $x^3+3x^2-2x$ **17.** $-a^3+6a^2+a$
**18.** $-2b^3-3b^2+6b$ **19.** $2x^4-3x^3-2x^2$ **20.** $-2y^5-3y^4+4y^3$ **21.** $-4y^4-10y^3+16y^2$
**22.** $12x^4-6x^3+21x^2$ **23.** $20x^4-4x^3-36x^2$ **24.** $-5y^4+15y^3+30y^2$ **25.** $a^3b-3a^2b^2-4ab^3$
**26.** $x^3y-2x^2y^2+2xy^3$ **27.** $a^3b+5a^2b^2-7ab^3$

## Objective 5.3.2

**1.** $x^3+4x^2+5x+2$ **2.** $x^3-4x^2+8x-5$ **3.** $a^3-3a^2+4x-4$ **4.** $x^3-7x^2+7x+15$
**5.** $a^3-6a^2+8a-3$ **6.** $3x^3-16x^2-7x+4$ **7.** $-2b^3+7b^2+6b-8$ **8.** $-2a^3+5a^2+7a-9$
**9.** $-2x^3+15x^2-24x+9$ **10.** $-3a^3-4a^2+5a+2$ **11.** $x^3+x^2-4x-4$ **12.** $y^3-9y$ **13.** $x^3-5x^2+3x-15$
**14.** $2y^2+2y^2-4y$ **15.** $x^4-3x^3-2x^2+7x-3$ **16.** $2y^4+y^3-6y^2-10y+15$ **17.** $8y^3-14y^2-51y+7$
**18.** $9y^3-3y^2-11y+6$ **19.** $3a^4-6a^3-2a^2+7a-6$ **20.** $4y^3+6y^2-8y+2$ **21.** $4a^4-12a^3-4a^2+13-3$
**22.** $6b^4-11b^3+3b^2+18b-6$ **23.** $y^4+2y^3-y^2+2$ **24.** $3a^4-8a^3+13a^2-9a+2$

## Objective 5.3.3

**1.** $y^2+5y+6$ **2.** $a^2+3a-10$ **3.** $b^2-b-20$ **4.** $y^2-5y-14$ **5.** $x^2+5x-36$ **6.** $y^2-8y+12$ **7.** $a^2-10a+16$
**8.** $x^2+8x-33$ **9.** $2x^2+13x+6$ **10.** $3y^2+5y+2$ **11.** $2x^2+3x-9$ **12.** $5x^2+13x-6$ **13.** $6x^2+13x+5$

**14.** $2y^2+7y-9$ **15.** $4y^2+y-14$ **16.** $30a^2+17a+2$ **17.** $12a^2-56a+65$
**18.** $10a^2+53a+63$ **19.** $15b^2+45b-44$ **20.** $12a^2+31a-30$ **21.** $x^2+3xy+2y^2$
**22.** $2a^2+5ab+2b^2$ **23.** $2x^2-5xy+3y^2$ **24.** $2a^2-3ab-9b^2$ **25.** $8a^2+18ab-5b^2$
**26.** $9x^2-9xy-10y^2$ **27.** $30x^2+17xy+2y^2$

## Objective 5.3.4

**1.** $x^2-1$ **2.** $x^2-9$ **3.** $x^2-49$ **4.** $4x^2-1$ **5.** $9x^2-1$ **6.** $x^2+10x+25$ **7.** $y^2-8y+16$
**8.** $9y^2-6y+1$ **9.** $x^2-2x+1$ **10.** $x^2+14x+49$ **11.** $x^2+18x+81$ **12.** $x^2-2xy+y^2$
**13.** $4a^2-20a+25$ **14.** $25x^2-40x+16$ **15.** $9x^2-42x+49$ **16.** $9a^2-25$ **17.** $4x^2+20x+25$
**18.** $81x^2-36x+4$ **19.** $a^2-4ab+4b^2$ **20.** $x^2+4xy+4y^2$ **21.** $25x^2-36$ **22.** $b^2-36a^2$
**23.** $x^2+10xy+25y^2$ **24.** $4-28y+49y^2$ **25.** $9-25y^2$ **26.** $16x^2-1$ **27.** $x^2+12xy+36y^2$

## Objective 5.3.5

**1.** $(9x^2-3x)$ m$^2$ **2.** $(3x^2-4x-4)$ in.$^2$ **3.** $(6x^2-x-1)$ ft$^2$ **4.** $(x^2+6x+9)$ cm$^2$
**5.** $(x^2-16x+64)$ km$^2$ **6.** $(4x^2+4x+1)$ in.$^2$ **7.** $(9x^2-24x+16)$ ft$^2$ **8.** $(\pi x^2-6\pi x+9\pi)$ m$^2$
**9.** $(\pi x^2+12\pi x+36\pi)$ in.$^2$ **10.** $(4\pi x^2+4\pi x+\pi)$ ft$^2$

## Objective 5.4.1

**1.** $\frac{1}{36}$ **2.** $\frac{1}{64}$ **3.** $\frac{1}{125}$ **4.** $\frac{1}{a^5}$ **5.** $\frac{x}{y^2}$ **6.** 1 **7.** $\frac{1}{a^{14}}$
**8.** $\frac{y^2}{x^2}$ **9.** $\frac{1}{x^8}$ **10.** 1 **11.** $3x$ **12.** $-3x^3$ **13.** $\frac{b^6}{a^2}$ **14.** $x^4y^3$
**15.** $y$ **16.** $5a$ **17.** $\frac{16}{a}$ **18.** $4a^2$ **19.** $\frac{-3a^2b^4}{4}$ **20.** $\frac{b^2}{-6a}$ **21.** $\frac{-2x^2y^2}{11z^3}$
**22.** $\frac{x^4y^3}{4}$ **23.** $-a^3b^2c$ **24.** $-\frac{4x^3}{3y^2}$

## Objective 5.4.2

**1.** $3.56\times10^{-5}$ **2.** $3.2\times10^{5}$ **3.** $3.7\times10^{-7}$ **4.** $5.1\times10^{-9}$ **5.** $6\times10^{10}$ **6.** $4.6\times10^{7}$ **7.** 0.00000214
**8.** 0.0000000053 **9.** 6,700,000,000 **10.** 5,230,000 **11.** 0.0045 **12.** 52,400,000 **13.** 0.028
**14.** 0.00000001638 **15.** 0.000000004 **16.** 70,000,000 **17.** 0.000000000018 **18.** $1.3\times10^{6}$ s
**19.** $4.2\times10^{6}$ mi

## Objective 5.5.1

**1.** $x+1$ **2.** $y+1$ **3.** $2a-3$ **4.** $2a-3$ **5.** $2b-5$ **6.** $2a+3$ **7.** $5y+3$
**8.** $3b^2-2$ **9.** $2y-1$ **10.** $x^2+2x-4$ **11.** $x^2-2x-1$ **12.** $a^3-6a^2-4a$ **13.** $xy-3$
**14.** $-2y^2+5$ **15.** $-4x^2+3$ **16.** $-y+4$ **17.** $-2+3x^2$ **18.** $4x+1-\frac{5}{x}$ **19.** $6y+4-\frac{5}{y}$

**20.** $-2x+5-\frac{4}{x}$ **21.** $4a-5+6b$ **22.** $2a+1-4b$ **23.** $5x+6-2y$ **24.** $b-4+3ab$

### Objective 5.5.2

**1.** $x+3$ **2.** $x+4$ **3.** $x-6$ **4.** $a-7$ **5.** $x-4$ **6.** $y+5$ **7.** $x+3$
**8.** $y-2$ **9.** $2x+3$ **10.** $2y-7$ **11.** $2y-5$ **12.** $2y-3$ **13.** $x-3$ **14.** $y+7$
**15.** $x-1$ **16.** $x-3+\frac{18}{x+3}$ **17.** $2x-2-\frac{6}{2x-3}$ **18.** $2y+2+\frac{2}{2y-1}$ **19.** $a-2+\frac{20}{a+5}$
**20.** $b-3-\frac{20}{b-4}$ **21.** $5y+3$ **22.** $6a-12+\frac{30}{2a+3}$ **23.** $2a+5$ **24.** $3x+4$

### Objective 5.5.3

**1.** $2x-5$ **2.** $4x-2+\frac{1}{x+3}$ **3.** $2x-7$ **4.** $3x+5+\frac{5}{x-4}$ **5.** $4x+4+\frac{1}{x-1}$
**6.** $4x-12+\frac{28}{x+3}$ **7.** $2x^2-x+7$ **8.** $3x^2+5x-3$ **9.** $3x^2+9x+28+\frac{69}{x-3}$ **10.** $x^2-6x+12-\frac{14}{x+2}$
**11.** $4x^2-x-2$ **12.** $3x^2+11x+5+\frac{7}{x-1}$ **13.** $2x^2+5x+19+\frac{105}{x-5}$ **14.** $4x^2-3x+1$ **15.** $3x^3-12x^2+28x-59+\frac{114}{x+2}$
**16.** $3x^3-x^4-4x+11+\frac{31}{x-3}$ **17.** $3x^3+6x^2+11x+22+\frac{47}{x-2}$ **18.** $x^3+3x^2+5x+15+\frac{27}{x-3}$

### Objective 5.5.4

**1.** 16 **2.** –25 **3.** –143 **4.** 193 **5.** 6 **6.** 19 **7.** 280
**8.** –4 **9.** 600 **10.** –19 **11.** 5 **12.** 4 **13.** 0 **14.** 579
**15.** 7 **16.** –2

## Chapter 6

### Objective 6.1.1

**1.** $4(a+1)$ **2.** $6(c-1)$ **3.** $4(2-a^2)$ **4.** $3(3+7x^2)$ **5.** $3(x+3)$ **6.** $5(2a^2+3)$ **7.** $8(3a-1)$
**8.** $12(2x+1)$ **9.** $3(3x-2)$ **10.** $4y(3x-4)$ **11.** $b^2(6-5b)$ **12.** $4x^2(5x-6)$ **13.** $12a^2(a^3-3)$
**14.** $6a^2b^2(4ab^2-3)$ **15.** $2ab(2a^4+3b^3)$ **16.** $5x^2y(5y-3x)$ **17.** $xy(3x-5y)$ **18.** $4a^2b^2(3a-3b)$
**19.** $x(x^2-5x+7)$ **20.** $4(x^2-4x+5)$ **21.** $3(2y^2-3y+4)$ **22.** $b^2(b^2-3b+7)$
**23.** $4x^2(1-2x+3x^2)$ **24.** $4(3y^2-4y+12)$ **25.** $5y(y^3+2y^2-7)$ **26.** $4x^2(x^2+3x-7)$
**27.** $15ab(3a^2b-5a^2+2b^3)$

### Objective 6.1.2

**1.** $(a+b)(x+3)$ **2.** $(x-y)(a+5)$ **3.** $(b-1)(x-y)$ **4.** $(a-1)(y-1)$ **5.** $(y+3)(a-1)$ **6.** $(y-7)(3x-1)$
**7.** $(x-5)(2x-1)$ **8.** $(x-3)(5y+1)$ **9.** $(a-2b)(x+y)$ **10.** $(x-9)(a-1)$ **11.** $(a-5)(b-2)$
**12.** $(x-3y)(c+d)$ **13.** $(a+1)(3x+4)$ **14.** $(a-4)(x-y)$ **15.** $(a-3b)(x-y)$ **16.** $(a-9)(x+1)$
**17.** $(x-7)(a-3)$ **18.** $(a-4)(2x+y)$ **19.** $(a-9)(m+n)$ **20.** $(n-2)(3m+1)$
**21.** $(b-1)(2a-c)$ **22.** $(a-2)(x-3y)$ **23.** $(a-b)(2-c)$ **24.** $(n-1)(m-2)$
**25.** $(c-7)(x+y)$ **26.** $(b-c)(a+3)$ **27.** $(x-y)(5+z)$

# Answers to AIM for Success Practice Sheets

## Objective 6.2.1

**1.** $(a-7)(a+5)$ **2.** $(a-3)(a-1)$ **3.** $(a+5)(a-1)$ **4.** $(b-5)(b-2)$ **5.** $(b+5)(b+3)$
**6.** $(b+8)(b-2)$ **7.** $(y-6)(y+5)$ **8.** $(y+11)(y-6)$ **9.** $(x-10)(x+6)$ **10.** $(y-2)(y-5)$
**11.** $(y-6)(y-3)$ **12.** $(x-6)(x-6)$ **13.** $(a+7)(a-4)$ **14.** $(x+8)(x+2)$
**15.** $(b-20)(b+9)$ **16.** $(x+5)(x+5)$ **17.** $(b+8)(b+2)$ **18.** $(x-12)(x+5)$
**19.** $(x+14)(x-4)$ **20.** $(x-16)(x+8)$ **21.** $(x-11)(x+7)$ **22.** $(b-14)(b-6)$
**23.** $(b-9)(b-12)$ **24.** $(b-12)(b-15)$ **25.** $(a+7)(a+9)$ **26.** $(x-15)(x-4)$ **27.** $(x-21)(x-4)$

## Objective 6.2.2

**1.** $2(x+3)(x+1)$ **2.** $3(x+4)(x+1)$ **3.** $3(a+4)(a-3)$ **4.** $4(x-3)(x+1)$
**5.** $a(b+7)(b-2)$ **6.** $x(y-2)(y-5)$ **7.** $x(y+7)(y+4)$ **8.** $x(x-6)(x-2)$
**9.** $3y(y-5)(y-1)$ **10.** $4(x+8)(x-1)$ **11.** $2(x-3)(x+2)$ **12.** $3(z-8)(z+1)$
**13.** $z(z+9)(z+5)$ **14.** $c(c+15)(c+2)$ **15.** $(x-7y)(x-2y)$ **16.** $4(y-8)(y+4)$
**17.** $3(x+3)(x-2)$ **18.** $2(x-10)(x+3)$ **19.** $3(y-5)(y-3)$ **20.** $5(y-1)(y-9)$
**21.** $2x(x+3)(x-2)$ **22.** $(y+8x)(y-x)$ **23.** $x(x+8)(x-5)$ **24.** $10x(x+2)(x+4)$
**25.** $(x^2+9)(x^2-7)$ **26.** $2(x+15)(x-8)$ **27.** $a^2(a+9)(a-8)$

## Objective 6.3.1

**1.** $(x-2)(2x-1)$ **2.** $(3x+1)(x-1)$ **3.** $(3x-2)(x+1)$ **4.** $(2b-1)(b-6)$ **5.** $(3a-1)(a-2)$
**6.** $(2x+3)(2x-1)$ **7.** $(5a-2)(a+3)$ **8.** $(3y-2)(2y+3)$ **9.** $(6x-5)(x+1)$
**10.** $(5x+2)(x-1)$ **11.** $(7y+1)(y+1)$ **12.** $(7x+1)(2x+1)$ **13.** $(7y+4)(y+2)$
**14.** $(3a+1)(3a-2)$ **15.** $(3a+4)(a-3)$ **16.** $(3x+2)(x-4)$ **17.** $(3x+2)(2x-3)$
**18.** $(4y+5)(y+5)$ **19.** $(7x-1)(x+3)$ **20.** $(5x+7)(x-1)$ **21.** $(5x+2)(2x-3)$
**22.** $(4x-3)(2x-5)$ **23.** $(4x-5)(3x+2)$ **24.** $(4x+5)(2x-3)$ **25.** $(5x+2)(2x-5)$
**26.** $(5x-2)(3x-4)$ **27.** $(6y+1)(2y-3)$

## Objective 6.3.2

**1.** $3(x-1)^2$ **2.** $2(4x-1)(x+3)$ **3.** $2(3y-7)(y-1)$ **4.** $5(3y+2)(y-1)$ **5.** $2b(a+6)(a-2)$
**6.** $(4x-y)(x-6y)$ **7.** $(x-6y)(x-5y)$ **8.** $x(2x+1)(x-4)$ **9.** $(5+x)(3+x)$ **10.** $(7-x)(3+x)$
**11.** $(8-x)(4+x)$ **12.** $(2y+1)(2y-3)$ **13.** $3(2y-1)(2y+3)$ **14.** $(5x-1)(3x-4)$
**15.** $5(x-8)(x+1)$ **16.** $x(5x-1)(x-1)$ **17.** $(3x+4)(x-5)$ **18.** $3(2y-3)(2y+1)$
**19.** $2(2x-1)(2x+3)$ **20.** $5a(2a-5)(a+1)$ **21.** $x(3x+1)(2x+5)$ **22.** $2x(3x-1)(x-4)$
**23.** $2(5b-3)(b+2)$ **24.** $z(11+z)(2-z)$ **25.** $2x(3x-1)(x-5)$ **26.** $(19-x)(1+x)$
**27.** $(13x-3)(x+4)$

## Objective 6.4.1

**1.** $-6, 4$ **2.** $-\frac{4}{3}, 7$ **3.** $0, \frac{8}{3}$ **4.** $-\frac{1}{3}, 6$ **5.** $0, 3$ **6.** $0, 3$ **7.** $-\frac{2}{5}, \frac{2}{5}$
**8.** $-3, 3$ **9.** $-3, 4$ **10.** $-3, 5$ **11.** $-2, -1, 0$ **12.** $-4, \frac{5}{3}$ **13.** $-1, 11$ **14.** $\frac{7}{3}, 2$
**15.** $-1, \frac{4}{3}$ **16.** $-\frac{3}{2}$

## Objective 6.4.2

**1.** –7, 6 **2.** –10, 9 **3.** length: 33 cm; width: 6 cm **4.** base: 8 cm; height: 22 cm **5.** 25 cm
**6.** length: 41 ft; width: 7 ft **7.** 3 s **8.** 1.5 s

## Objective 6.4.3

**1. 2. 3. 4.** $(xy-10)(xy+2)$ $(xy-8)(xy-5)$ $(ab+7)(ab+4)$ $(ab-7)(ab-8)$

**5. 6. 7. 8.** $(x^2-8)(x^2-2)$ $(y^2-14)(y^2+3)$ $(b^2-7)(b^2-3)$ $(b^2+9)(b^2+3)$

**9. 10. 11.** Nonfactorable **12.** $(a^2-12)(a^2+5)$ $(a^2b^2-2)(a^2b^2+12)$ $(a^n-5)(a^n+3)$

**13. 14. 15. 16.** $(2xy-5)(xy+4)$ $(3xy-4)(xy+2)$ $(3ab-2)(ab-5)$ $(2x^n+3)(x^n+3)$

**17. 18.** $(2x^n+1)(x^{n-5})$ $(3a^n+2)(2a^n+1)$

## Objective 6.5.1

**1.** $3(x+2)(x-2)$ **2.** $2(x+5)(x-5)$ **3.** $x(x+1)^2$ **4.** $y(y-4)^2$ **5.** $x^2(x+3)(x-2)$
**6.** $a^2(a-8)(a+5)$ **7.** $3(b+3)(b+7)$ **8.** $4(y-1)(y-7)$ **9.** $2(a-11)(a+2)$
**10.** $x(x-10)(x+2)$ **11.** $b(b-6)(b+1)$ **12.** $5a(a-3)^2$ **13.** $2(2x-1)(x-1)$ **14.** $x^2(2x-1)(x-5)$
**15.** $x^2(x+4)(x-4)$ **16.** $3(5x-1)(x-1)$ **17.** $3a(3a+1)(a+4)$ **18.** $x(3y-4)(y+5)$
**19.** $3(4-x)(2+x)$ **20.** $b^2(a+8)(a-1)$ **21.** $4y(x+2)(x+1)$ **22.** $18a(a-2)(a-1)$
**23.** $y^2(x+1)(x-1)$ **24.** $5(x+3y)(x-3y)$ **25.** $x^2(x+3)(x-3)$ **26.** $(3x-3)(a+b)$
**27.** $(a+b)(y+2)(y-2)$

## Objective 6.5.1

**1. 2. 3. 4. 5.** $4(x-1)^2 5(x+2)(x-2)$ $6(x+1)(x-1)$ **4.**$4(2x+3)(2x-3)$ $(x^2+9)(x+3)(x-3)$

**6.** Nonfactorable **7. 8. 9.** $3x^2(2x+5)(2x-5)$ $3a(1-a)(1+a+a^2)a^3(ab+3)(a^2b^2-3ab+9)$
**10. 11. 12. 13. 14.**
$3x^2(x+8)(x-5)$ $x^2(x-7)(x+4)$ $(a^2+9)(a+2)(a-2)$ **1**$(a^2+1)(a+3)(a-3)$ $3a^2(a-4)^2$

**15. 16. 17. 18.** $(a+1)(a-1)(a+2)(a-2)$ $(x^2+4)(x+2)(x-2)$ **1**$5(2x-3)(x+1)$ $6(y+3)(y+2)$

**19. 20. 21. 22. 23.** $3xy(x-2y)(5x+4y)$ **20**$2x^2(2x+3y)(4x-5y)$ $a^2(a^n-4)^2$ **22.** $x(x(x^n+2)^2x^n(x-2)(2x-3)$

**24.** $a^n(2a-1)(a+3)$

## Objective 6.6.1

**1.** –2, –1 **2.** 4, 6 **3.** 1, 6 **4.** –7, 5 **5.** 0, 8 **6.** –1, 0 **7.** 0, 4
**8.** $-\frac{2}{3}, 0$ **9.** $0, \frac{5}{2}$ **10.** $0, \frac{1}{2}$ **11.** $-\frac{3}{4}, 0$ **12.** –1, 4 **13.** –4, 7 **14.** –3, 7
**15.** –8, 2 **16.** –1, 6 **17.** –2, 7 **18.** $-1, \frac{3}{2}$ **19.** $\frac{1}{4}, 3$ **20.** $-3, \frac{2}{5}$ **21.** $-2, -\frac{1}{2}$
**22.** –11, 1 **23.** 3, 6 **24.** –10, 5 **25.** $-4, \frac{7}{2}$ **26.** –4, 12 **27.** $-4, \frac{3}{2}$

## Objective 6.6.2

**1.** 4 **2.** –6 **3.** 2, 5 **4.** 4, 6 **5.** 3, 9 **6.** 3 **7.** 5, 6
**8.** 10, 12 **9.** base: 24 ft; height: 8 ft **10.** length: 24 in.; width: 6 in. **11.** length: 15 ft; width: 8 ft **12.** 8 in.

# CHAPTER 7

## Objective 6.1.1

**1.** $-\frac{7}{2}$ **2.** $\frac{2}{17}$ **3.** $-\frac{1}{2}$ **4.** $-\frac{2}{9}$ **5.** $\frac{3}{29}$ **6.** $\frac{6}{5}$ **7.** $-\frac{5}{19}$
**8.** $-\frac{1}{3}$ **9.** $\{x \mid x \neq -3\}$ **10.** $\{x \mid x \neq -3, 4\}$ **11.** $\{x \mid x \neq -1\}$ **12.** $\{x \mid x \neq 7\}$ **13.** $\{x \mid x \neq -3, 3\}$
**14.** $\{x \mid x \neq -2, 5\}$ **15.** $\{x \mid x \neq -4\}$ **16.** $\{x \mid x \neq -2, 3\}$

## Objective 7.1.2

**1.** $1+3x$ **2.** $3x-1$ **3.** $5x$ **4.** $\frac{4x}{3}$ **5.** $-\frac{3}{x}$ **6.** $2x^2-x+3$ **7.** $\frac{x^n-3}{4}$
**8.** $\frac{a-3}{a+7}$ **9.** $\frac{x-5}{x+5}$ **10.** $\frac{x^n+y^n}{x^n-y^n}$ **11.** $\frac{3}{a^n-3}$ **12.** $-\frac{x+4}{2x-1}$ **13.** $\frac{3x+2}{3x-2}$ **14.** $\frac{x-4y}{x+4y}$
**15.** $\frac{x+9}{x-3}$ **16.** $\frac{9x^2-3xy+y^2}{3x+y}$ **17.** $\frac{2xy+1}{2xy+5}$ **18.** $\frac{x-3}{a-b}$ **19.** $\frac{x^2-8}{x^2+8}$ **20.** $\frac{x^2+4}{x^2+3}$
**21.** $\frac{xy-4}{xy-7}$ **22.** $\frac{a^n+4}{a^n+6}$ **23.** $\frac{a^n+2}{a^n+5}$

## Objective 7.2.1

**1.** $\frac{ab^2}{x}$ **2.** $\frac{2x-1}{y^2(x+1)}$ **3.** $\frac{x-3}{x-1}$ **4.** $\frac{x+4}{x+2}$ **5.** $\frac{x+6}{x-6}$ **6.** $-\frac{2x-5}{2x+5}$ **7.** $\frac{4(x+3)}{3x(x-1)}$
**8.** $x+2y$ **9.** $\frac{(x-1)(3x-2)}{(3x+1)(x+5)}$ **10.** $\frac{y}{8ax^2}$ **11.** $\frac{14}{15}$ **12.** $\frac{x^2y(2x+1)}{x-1}$ **13.** $\frac{x-2}{x+6}$
**14.** 2 **15.** $\frac{x-2}{x-4}$ **16.** $\frac{2(x+3)}{x-2}$ **17.** 2 **18.** $\frac{(x^n+3)^2}{(x^n-3)(x^n-2)}$

## Objective 7.2.2

**1.** $\frac{3x^2-9x}{6x^2(x-3)}, \frac{2x-6}{6x^2(x-3)}$ **2.** $\frac{2x^3-3x^2}{4x^3(2x-1)}, \frac{6x-3}{4x^3(2x-1)}$ **3.** $\frac{3x-2}{4x^2-12x}, \frac{-16x^3+48x^2}{4x^2-12x}$
**4.** $\frac{5x-4}{2x(x-4)}, \frac{6x^3-24x^2}{2x(x-4)}$ **5.** $\frac{30x-18}{(5x+3)(5x-3)}, \frac{-40x-24}{(5x+3)(5x-3)}$ **6.** $\frac{5x}{(x+2)(x-2)}, \frac{x^2+5x+6}{(x+2)(x-2)}$
**7.** $\frac{6x}{3(3+x)(3-x)}, \frac{12x+x^2}{3(3+x)(3-x)}$ **8.** $\frac{15}{6(x+2y)(x-2y)}, \frac{6x+12y}{6(x+2y)(x-2y)}$ **9.** $\frac{15x}{5(x+5)(x-5)}, \frac{x^2+3x-10}{5(x+5)(x-5)}$
**10.** $\frac{2x^2+2x}{(x+1)^2(x-1)}, \frac{3x^3-3x}{(x+1)^2(x-1)}$ **11.** $\frac{x^2-1}{(x-2)(x^2+2x+4)}, \frac{3x-6}{(x-2)(x^2+2x+4)}$

**12.** $\frac{x-4}{(4-x)(16+4x+x^2)}, \frac{12-3x}{(4-x)(16+4x+x^2)}$ **13.** $\frac{7x^2-14x}{(x+3)(x-2)^2}, \frac{-4x^2-12x}{(x+3)(x-2)^2}$

**14.** $\frac{2x^2+6x}{(x-1)(x+3)^2}, \frac{3x^2-3x}{(x-1)(x+3)^2}$ **15.** $\frac{-3x^2+9x}{(2x+1)(x-2)(x-3)}, \frac{2x^2-4x}{(2x+1)(x-2)(x-3)}$

**16.** $\frac{10x^2+25}{(2x-5)(2x+5)(2x+3)}, \frac{-6x^2+15x}{(2x-5)(2x+5)(2x+3)}$ **17.** $\frac{4}{(2x-1)(x+3)}, \frac{-3x^2-9x}{(2x-1)(x+3)}, \frac{4x^2-1}{(2x-1)(x+3)}$

**18.** $\frac{6}{(3x-2)(x-4)}, \frac{-3x^2-2x}{(3x-2)(x-4)}, \frac{x^2-5x+4}{(3x-2)(x-4)}$ **19.** $\frac{2x^2+4x}{(x-5)(x+2)}, \frac{4x-20}{(x-5)(x+2)}, \frac{-x-3}{(x-5)(x+2)}$

**20.** $\frac{3x^2+9x}{(x-2)(x+3)}, \frac{-x+2}{(x-2)(x+3)}, \frac{-x+2}{(x-2)(x+3)}$ **21.** $\frac{3x^n+6}{(x^n+2)^2(x^n-2)}, \frac{6x^n-12}{(x^n+2)^2(x^n-2)}$

**22.** $\frac{x-3}{(x^n+2)(x^n-1)}, \frac{5x^{n+1}+10x}{(x^n+2)(x^n-1)}$

## Objective 7.2.3

**1.** 0 **2.** $\frac{1}{5x^2}$ **3.** $\frac{1}{x+1}$ **4.** $\frac{1}{x+5}$ **5.** $\frac{1}{x+y}$ **6.** $\frac{1}{x+2}$ **7.** $\frac{5-4xy-35y}{15xy^2}$

**8.** $\frac{18b-15a-2}{24a^2b^2}$ **9.** $\frac{11}{30x}$ **10.** $\frac{a^2+10a-4}{a(a-2)}$ **11.** $\frac{2(x^2+8x+2)}{(x+5)(x+4)}$ **12.** $\frac{10-2x}{(x+5)^2}$

**13.** $\frac{x+1}{x+5}$ **14.** $\frac{-2x+4}{x^2+8x+16}$ **15.** $\frac{x^2-9x}{(x-4)(x-5)}$ **16.** $\frac{-5a^2+38a}{(a+4)(a-1)}$ **17.** $\frac{4}{(x+4)(x+3)}$

**18.** $-\frac{x^{2n}+x^n+9}{(x^n-3)(x^n+4)}$ **19.** $\frac{5x+14}{9-4x^2}$ **20.** $\frac{x^2-40x+83}{5x^2-20}$

## Objective 7.3.1

**1.** $\frac{x}{3x-1}$ **2.** $\frac{4-x}{x}$ **3.** $-\frac{a}{a+5}$ **4.** $\frac{6-a}{a}$ **5.** 1 **6.** $\frac{x}{4(4-x)}$ **7.** $\frac{x+4}{x+6}$

**8.** $\frac{3-2x}{2-3x}$ **9.** $\frac{3x-1}{x-3}$ **10.** $\frac{2(x+8)}{5x-3}$ **11.** $\frac{2x-8}{2x^2-3x-24}$ **12.** $\frac{3}{3x-4}$ **13.** $\frac{x-4}{x+6}$

**14.** $\frac{x-1}{x+2}$ **15.** $-\frac{2(a+3)}{4a-3}$ **16.** $-\frac{3(a-4)}{8(a+2)}$ **17.** $-\frac{4x}{x^2+4}$ **18.** $-\frac{4}{x}$

## Objective 7.4.1

**1.** –3 **2.** $\frac{5}{2}$ **3.** 1 **4.** –3 **5.** –2 **6.** –4 **7.** 3
**8.** –2 **9.** –5 **10.** 12 **11.** 5 **12.** no solution **13.** –3 **14.** 2
**15.** –6 **16.** –9 **17.** 4 **18.** –8

## Objective 7.4.2

**1.** 36 min **2.** 18 min **3.** 12 min **4.** 12 h **5.** 24 h **6.** 30 h **7.** 6 min
**8.** 5 h **9.** 24 min **10.** 4.8 h

## Objective 7.4.3

**1.** 48 mph **2.** corporate jet: 435 mph; commercial jet: 540 mph **3.** 50 mph **4.** 400 mph **5.** 50 mph

**6.** 55 mph **7.** 28 mph **8.** 60 mph **9.** freight train: 40 mph; passenger train: 55 mph **10.** 60 mph

### Objective 7.5.1

**1.** 6 **2.** 4 **3.** 7 **4.** –3 **5.** $-\frac{5}{3}$ **6.** 6 **8.** 5

**9.** 4 **10.** \$90,000 **11.** \$115.20 **12.** 76 ft by 96 ft **13.** \$93.75 **14.** $\frac{5}{8}$ oz

**15.** 15 oz

### Objective 7.5.2

**1.** 21 lb **2.** 10 lb/in$^2$ **3.** 234.4 ft **4.** 128 ft **5.** 15 rpm **6.** 105 lb/in$^2$ **7.** 192 lumens
**8.** 16 ft **9.** 135 times/min **10.** 17.5 amps

### Objective 7.6.1

**1.** $n=\frac{a_n-a_1}{d}+1$ **2.** $b=\frac{2A}{h}$ **3.** $a=-\frac{b}{2a}$ **4.** $n=\frac{PV}{RT}$ **5.** $G=\frac{Fr^2}{m_1m_2}$ **6.** $V_2=\frac{P_1V_1T_2}{P_2T_1}$

**7.** $b_1=\frac{2A-hb_2}{h}$ **8.** $r=\frac{D}{t}$ **9.** $C=\frac{100M}{I}$ **10.** $T=\frac{E+alt}{al}$ **11.** $a=\frac{bt}{b-t}$ **12.** $v=\frac{2at-2d}{a}$

**13.** $g=\frac{2h-2vt}{t^2}$ **14.** $C=Rt+S$ **15.** $m=\frac{RK+T}{B}$ **16.** $m=\frac{K}{r-v}$ **17.** $I=\frac{\pi r^2-S}{\pi r}$ **18.** $c=\frac{A-P}{A}$

## CHAPTER 8

### Objective 8.1.1

**1.** 5 **2.** 3 **3.** 64 **4.** $\frac{1}{4}$ **5.** Not a real number **6.** –3 **7.** $\frac{7}{4}$

**8.** $\frac{1}{x^{2/5}}$ **9.** $a^{5/3}$ **10.** $x^{3/4}$ **11.** $a^{4n}$ **12.** $a$ **13.** $\frac{1}{x^{1/5}}$ **14.** $a^3b^6$

**15.** $\frac{x^2}{y^3}$ **16.** $x^2y^{12}$ **17.** $x^{4/3}-x^{1/3}$ **18.** $a$ **19.** $x^{2n}$ **20.** $a^{n/4}$ **21.** $\frac{1}{x^{5/3}}$

**22.** $\frac{1}{a^{2n}}$ **23.** $x$ **24.** $\frac{1}{x}$ **25.** $\frac{1}{a^4}$ **26.** $\frac{1}{x^6}$ **27.** $\frac{1}{x^9}$

### Objective 8.1.2

**1.** $\sqrt[3]{2}$ **2.** $\sqrt{7}$ **3.** $\sqrt[5]{b^3}$ **4.** $\sqrt[4]{128x^7}$ **5.** $-9\sqrt[3]{a^2}$ **6.** $\sqrt[3]{ab^2}$ **7.** $\sqrt[5]{(x^3y^2)^3}$

**8.** $\sqrt[3]{(a^2b^2)^2}$ **9.** $\sqrt[3]{3a-2b}$ **10.** $\sqrt[3]{3x-4}$ **11.** $\sqrt[3]{(2x-1)^2}$ **12.** $\frac{1}{\sqrt[4]{x^3}}$ **13.** $15^{1/2}$ **14.** $y^{1/3}$

**15.** $(5y^5)^{1/4}$ **16.** $-(2x^3)^{1/2}$ **17.** $-(3x^5)^{1/3}$ **18.** $-(2x^7)^{1/4}$ **19.** $-(4x^4)^{1/3}$ **20.** $2xy^{2/3}$ **21.** $3xy^{1/4}$

**22.** $(a^2-8)^{1/2}$ **23.** $(2-x^2)^{1/2}$ **24.** $(x^2+y^2)^{1/2}$

### Objective 8.1.3

**1.** $x^6$ **2.** $x^5$ **3.** $-a^2$ **4.** $x^2y^3$ **5.** $-x^4y^2$ **6.** $-a^2b^5$ **7.** $-ab^3$

**8.** $x^2y^5$ **9.** $x^3y^5$ **10.** $-a^3b^4$ **11.** $12x^6y^8$ **12.** $3a^2b^5$ **13.** $9x^3y^2$ **14.** $4x^2$

**15.** $5x^4y$ **16.** $-x^4y^7$ **17.** $-2a^6b^3$ **18.** $-a^2b^3$ **19.** $x^4y^6$ **20.** $x^3y^5$ **21.** $2a^4b^5$
**22.** $5a^2b^3$ **23.** $-ab^3$ **24.** $3x^2y^3$

## Objective 8.2.1

**1.** $x^2y^3z^3\sqrt{z}$ **2.** $7a^2b^3\sqrt{2}$ **3.** $5y^3z^4\sqrt{2xz}$ **4.** $5xy^2z^3\sqrt{3x}$ **5.** Not a real number **6.** $-xy^3\sqrt{x}$
**7.** $x^2y^3z^4\sqrt{xyz}$ **8.** $x^3y^4z^2$ **9.** $x^3z^6$ **10.** $8y^3z^5\sqrt{3xy}$ **11.** $8xy^2z^3\sqrt{2xy}$ **12.** $2x^2y^3z^3\sqrt[3]{z}$
**13.** $-4xy^3$ **14.** Not a real number **15.** Not a real number **16.** $ab^2\sqrt[3]{a^2b}$ **17.** $a^2b^2\sqrt[3]{b^2}$ **18.** $-2x^2y\sqrt[3]{y}$
**19.** $-4x^2y^3$ **20.** $ab^2c^2\sqrt[3]{a^2c}$ **21.** $a^4b^3c^2\sqrt[3]{c^2}$ **22.** $a^3b^2c^2\sqrt[3]{b}$ **23.** $2xy\sqrt[4]{x^3y}$ **24.** $2x^3y^2\sqrt[4]{y}$

## Objective 8.2.2

**1.** $-4\sqrt{x}$ **2.** $14\sqrt{x}$ **3.** $-\sqrt{5}$ **4.** $\sqrt{3}$ **5.** $\sqrt{2x}$ **6.** $11\sqrt{3x}$ **7.** $\sqrt{6a}$
**8.** $9\sqrt{3a}$ **9.** $13x\sqrt{3x}$ **10.** $2x\sqrt{3xy}$ **11.** $-7a\sqrt[3]{2a^2}$ **12.** $14a^2b^2\sqrt{2ab}$ **13.** $3ab^4\sqrt{ab}$
**14.** $-1$ **15.** $6\sqrt[3]{3}$ **16.** $-9a\sqrt[3]{2a}$ **17.** $20y\sqrt{3y}$ **18.** $6a\sqrt{a}$ **19.** $ab\sqrt[4]{ab}$ **20.** $6xy\sqrt[3]{2x}$
**21.** $-13a^2\sqrt{2a}$ **22.** $-7x^2y\sqrt[4]{3xy}$

## Objective 8.2.3

**1.** 4 **2.** $4\sqrt{3}$ **3.** 14 **4.** $4\sqrt[3]{4}$ **5.** $x^2y^4$ **6.** $3x^2y^3\sqrt[3]{2}$ **7.** $4ab^3\sqrt[3]{2a}$
**8.** 8 **9.** $-4$ **10.** $x+\sqrt{3x}$ **11.** $x-\sqrt{2x}$ **12.** $8x\sqrt{2}-6\sqrt{2x}$ **13.** $2x+6\sqrt{2x}+9$
**14.** $x+4\sqrt{x}+4$ **15.** $3x-12\sqrt{3x}+36$ **16.** $48x^3y^3\sqrt{3}$ **17.** $900x^3y^4$ **18.** $6a^2b^2$ **19.** $3a^2b^3\sqrt[3]{6a^2}$
**20.** $\sqrt{2}$ **21.** $-2\sqrt{3}$ **22.** $x-9$ **23.** $x-16$

## Objective 8.2.4

**1.** $4x$ **2.** $a^2b\sqrt{2}$ **3.** $3ab^2$ **4.** $\frac{\sqrt{3}}{3}$ **5.** $\frac{\sqrt{5}}{5}$ **6.** $\frac{\sqrt{6x}}{6x}$ **7.** $\frac{3\sqrt{2y}}{2y}$
**8.** $\frac{3\sqrt{x}}{x}$ **9.** $\frac{2\sqrt{6a}}{3a}$ **10.** $\frac{\sqrt{5x}}{x}$ **11.** $\frac{\sqrt{2a}}{a}$ **12.** $\frac{2\sqrt[3]{9}}{3}$ **13.** $\sqrt[3]{3}$ **14.** $\frac{x\sqrt{2xy}}{2y}$
**15.** $\frac{b\sqrt{6}}{4a^2}$ **16.** $\frac{x^3\sqrt{15}}{5}$ **17.** $\frac{3\sqrt{5}-3}{4}$ **18.** $\frac{9}{4}+\frac{3}{4}\sqrt{5}$ **19.** $\frac{2\sqrt{x}+8}{x-16}$ **20.** $\frac{5\sqrt{x}+25}{25-x}$ **21.** $\frac{7-2\sqrt{10}}{3}$
**22.** 11 **23.** 37 **24.** $x+1$ **25.** $x+1$ **26.** $\frac{9}{2}x\sqrt{2}-12x$ **27.** $2x-4$

## Objective 8.3.1

**1.** Domain: $\{x|x\ge 5\}$ **2.** Domain: $\{x|x\le 4\}$ **3.** Domain: $\{x|x\ge 0\}$ **4.** Domain: $\{x|x\ge 0\}$
**5.** Domain: $\{x|x\in \text{ real numbers}\}$ **6.** Domain: $\{x|x\ge 0\}$ **7.** Domain: $(-\infty,\infty)$ **8.** Domain: $[0,\infty)$
**9.** Domain: $(-\infty,1]$ **10.** Domain: $[-2,\infty)$ **11.** Domain: $(-\infty,\infty)$ **12.** $(-\infty,-3]$

## Objective 8.3.2

**1.** 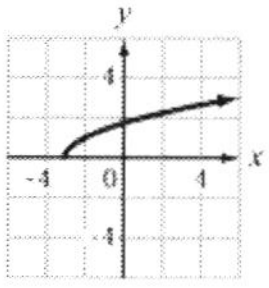


**2.** 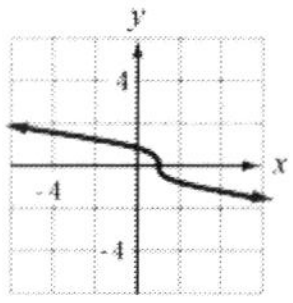


**3.** 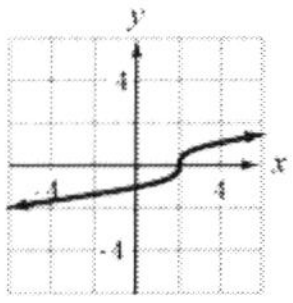


**4.** 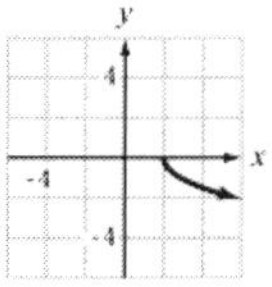


**5.** 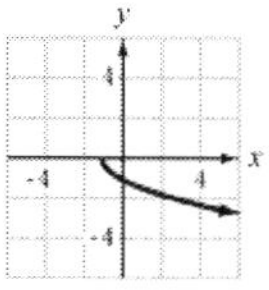


**6.** 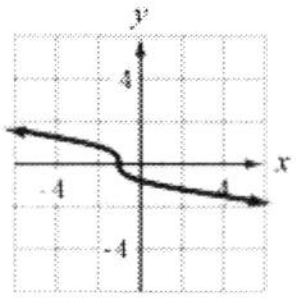


**7.** 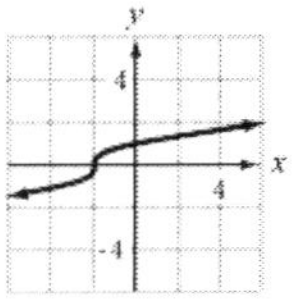


**8.** 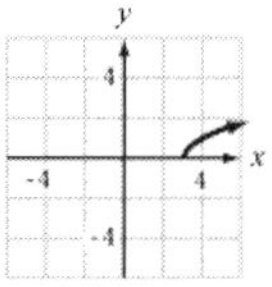


## Objective 8.4.1

**1.** 9 **2.** 49 **3.** 36 **4.** 8 **5.** 216 **6.** 12 **7.** –72
**8.** No solution **9.** 16 **10.** 3 **11.** 6 **12.** 4 **13.** 20 **14.** 8
**15.** –7 **16.** 7 **17.** 2 **18.** 0

## Objective 8.4.2

**1.** 12 ft **2.** 18.7 ft **3.** 12.25 ft **4.** 14.88 ft **5.** 87.89 ft **6.** 1600 ft **7.** 152 m
**8.** 200 ft **9.** 3.25 ft **10.** 8.31 ft

## Objective 8.5.1

**1.** $7i$ **2.** $2i\sqrt{2}$ **3.** $2xi\sqrt{3}$ **4.** $10i\sqrt{2}$ **5.** $8i\sqrt{2}$ **6.** $12i$ **7.** $6i\sqrt{3x}$
**8.** $6yi\sqrt{5y}$ **9.** $3a^3+3a^3i$ **10.** $4+5i$ **11.** $6+4i$ **12.** $7xy^2+10x^2yi$ **13.** $4\sqrt{3}-3i\sqrt{3}$
**14.** $5y\sqrt{2x}+3xi\sqrt{2y}$ **15.** $2x^2y^2z\sqrt{6xz}-6xy^4z^3i$ **16.** $5\sqrt{2}-7i\sqrt{2}$ **17.** $5\sqrt{6}-3i\sqrt{3}$
**18.** $6x^2z\sqrt{2x}-8y^3i\sqrt{2}$

## Objective 8.5.2

**1.** $11-6i$ **2.** $11-9i$ **3.** $12+3i$ **4.** $12-i$ **5.** $13-7i$ **6.** $-6-2i$ **7.** $-6+i$
**8.** $1-7i\sqrt{3}$ **9.** $9\sqrt{2}-4i\sqrt{2}$ **10.** $-2\sqrt{6}-5i\sqrt{5}$ **11.** $9-i$ **12.** $8-3i$ **13.** $6-i$
**14.** $7-i$ **15.** $0$ **16.** $-3-3i$ **17.** $2-i$ **18.** $5+2i$

## Objective 8.5.3

**1.** $40$ **2.** $-15$ **3.** $-160i$ **4.** $-32$ **5.** $-9$ **6.** $-8$ **7.** $-4\sqrt{3}$
**8.** $-3+15i$ **9.** $-12+8i$ **10.** $17-i$ **11.** $9+13i$ **12.** $24-10i$ **13.** $9-7i$ **14.** $2-11i$
**15.** $40$ **16.** $34$ **17.** $13-i$ **18.** $3-21i$ **19.** $6+13i$ **20.** $21+10i$ **21.** $5$
**22.** $\frac{17}{4}i$ **23.** $36+3i$ **24.** $\frac{5}{3}+\frac{5}{3}i$

## Objective 8.5.4

**1.** $-4i$ **2.** $-\frac{2}{3}i$ **3.** $-\frac{3}{4}i$ **4.** $2i+3$ **5.** $3i-\frac{2}{5}$ **6.** $\frac{20}{17}+\frac{5}{17}i$ **7.** $\frac{12}{5}-\frac{4}{5}i$
**8.** $\frac{25}{29}-\frac{10}{29}i$ **9.** $\frac{16}{17}+\frac{4}{17}i$ **10.** $\frac{15}{13}+\frac{3}{13}i$ **11.** $-i$ **12.** $-i$ **13.** $\frac{18}{17}+\frac{21}{17}i$ **14.** $\frac{20}{13}+\frac{48}{13}i$
**15.** $\frac{21}{10}-\frac{3}{10}i$ **16.** $4i$ **17.** $-\frac{2\sqrt{2}}{3}+\frac{2}{3}i$ **18.** $-\frac{1}{5}+\frac{2}{5}i$ **19.** $-\frac{\sqrt{2}}{6}+\frac{1}{3}i$ **20.** $\frac{15}{26}-\frac{29}{26}i$ **21.** $\frac{4}{5}-\frac{7}{5}i$

**22.** $1+2i$ **23.** $-\frac{3}{2}+\frac{7}{2}i$ **24.** $\frac{7}{5}+\frac{16}{5}i$

# CHAPTER 9

## Objective 9.1.1

**1.** –5 and 0 **2.** –4 and 4 **3.** –8 and 8 **4.** –4 and 3 **5.** 2 **6.** –4 **7.** 0 and 3
**8.** –5 and 0 **9.** –3 and 5 **10.** –4 and 1 **11.** 2 and 6 **12.** –1 and 12 **13.** –1 and $\frac{2}{3}$ **14.** $-\frac{1}{2}$ and 3
**15.** $-\frac{2}{3}$ and 3 **16.** and 2 **17.** –2 and 6 **18.** –2 and 1 **19.** –5 and 3 **20.** –6 and 2 **21.** $-7b$ and $3b$
**22.** $b$ and $4b$ **23.** $-c$ and $8c$ **24.** $-3a$ and $4a$ **25.** $\frac{c}{2}$ and $2c$ **26.** $-3a$ and $\frac{2a}{3}$

## Objective 9.1.2

**1.** $x^2-5x+4=0$ **2.** $x^2-5x+6=0$ **3.** $x^2+3x+2=0$ **4.** $x^2-2x-8=0$
**5.** $x^2+8x+15=0$ **6.** $x^2-4=0$ **7.** $x^2-16=0$ **8.** $x^2-6x+9=0$ **9.** $x^2-7x+6=0$
**10.** $x^2-7x=0$ **11.** $2x^2+3x=0$ **12.** $x^2-x=0$ **13.** $3x^2-7x+2=0$ **14.** $2x^2-7x+3=0$
**15.** $4x^2-7x-2=0$ **16.** $3x^2-13x+4=0$ **17.** $2x^2-3x-9=0$ **18.** $4x^2-x-3=0$
**19.** $3x^2-11x-4=0$ **20.** $4x^2-x-3=0$ **21.** $3x^2-4x+1=0$ **22.** $15x^2-19x+6=0$
**23.** $10x^2-3x-4=0$ **24.** $8x^2-10x-3=0$ **25.** $6x^2+5x+1=0$ **26.** $12x^2+5x-3=0$
**27.** $6x^2-7x-5=0$

## Objective 9.1.3

**1.** –5 and 5 **2.** –6 and 6 **3.** $-3i$ and $3i$ **4.** $-7i$ and $7i$ **5.** –1 and 1 **6.** –3 and 3 **7.** $-5i$ and $5i$
**8.** $-3\sqrt{3}$ and $3\sqrt{3}$ **9.** $-5i\sqrt{2}$ and $5i\sqrt{2}$ **10.** –7 or 1 **11.** and 6 **12.** –5 or 1 **13.** –7 or 11
**14.** –4 and 6 **15.** $\frac{2}{3}$ and 0 **16.** –1 and 0 **17.** $5-i\sqrt{15}$ and $5+i\sqrt{15}$ **18.** $-4-2i\sqrt{7}$ and $-4+2i\sqrt{7}$
**19.** $1-3\sqrt{2}$ and $1+3\sqrt{2}$ **20.** $-2-4i$ and $-2+4i$ **21.** $3-5\sqrt{2}$ and $3+5\sqrt{2}$ **22.** $\frac{3}{5}+5\sqrt{3}$ and $\frac{3}{5}-5\sqrt{3}$
**23.** $\frac{1}{3}-3\sqrt{3}$ and $\frac{1}{3}+3\sqrt{3}$ **24.** $-\frac{1}{2}-3\sqrt{5}$ and $-\frac{1}{2}+3\sqrt{5}$

## Objective 9.2.1

**1.** –7 and 1 **2.** –3 and 5 **3.** 2 **4.** –4 **5.** 1 and 5 **6.** $-4-\sqrt{6}$ and $-4+\sqrt{6}$
**7.** –1 and 3 **8.** –5 and 2 **9.** –4 and 9 **10.** 3 and 4 **11.** 3 and 6 **12.** $-2-2\sqrt{5}$ and $-2+2\sqrt{5}$
**13.** $3-\sqrt{7}$ and $3+\sqrt{7}$ **14.** –1 and $\frac{1}{3}$ **15.** $\frac{1}{2}$ and 1 **16.** $-\frac{3}{2}$ and 4 **17.** –2 and 12 **18.** –2 and 8
**19.** –1 and 5 **20.** $\frac{2}{5}$ and $\frac{1}{2}$ **21.** $\frac{1}{2}$ and $\frac{3}{2}$ **22.** $-\frac{2}{3}$ and 6 **23.** $\frac{1}{3}$ and $-\frac{3}{2}$ **24.** $-\frac{1}{2}$ and $\frac{7}{2}$ **25.** $-\frac{5}{2}$ and 2
**26.** $\frac{-1-\sqrt{5}}{2}$ and $\frac{-1+\sqrt{5}}{2}$ **27.** 2 and 5

## Objective 9.2.2

**1.** −5 and 10 **2.** −5 and 8 **3.** −9 and 1 **4.** $\frac{5}{4}-\frac{\sqrt{23}}{4}i$ and $\frac{5}{4}+\frac{\sqrt{23}}{4}i$ **5.** $\frac{1}{4}$ and $-\frac{3}{2}$ **6.** $\frac{1}{3}$ and $-\frac{3}{2}$
**7.** $\frac{1}{2}$ and 1 **8.** $-\frac{1}{2}$ and $-\frac{9}{2}$ **9.** $-\frac{1}{2}-\frac{3}{2}i$ and $-\frac{1}{2}+\frac{3}{2}i$ **10.** $2-i\sqrt{2}$ and $2+i\sqrt{2}$
**11.** $1-\frac{2}{3}\sqrt{3}$ and $1+\frac{2}{3}\sqrt{3}$ **12.** $-\frac{1}{3}$ and $\frac{3}{4}$ **13.** $\frac{1}{2}$ and 5 **14.** $\frac{11}{12}-\frac{\sqrt{95}}{12}i$ and $\frac{11}{12}+\frac{\sqrt{95}}{12}i$
**15.** $\frac{1}{2}-\frac{\sqrt{39}}{6}i$ and $\frac{1}{2}+\frac{\sqrt{39}}{6}i$ **16.** −3 and $\frac{4}{5}$ **17.** 0 and $\frac{5}{2}$ **18.** $-\frac{4}{3}$ and 3 **19.** 0.354 and 5.646
**20.** 0.394 and 7.606 **21.** 0.628 and 6.372 **22.** one real **23.** two real **24.** two complex

## Objective 9.3.1

**1.** −2, −1, 1, 2 **2.** −4, −2, 2, 4 **3.** −4, −1, 1, 4 **4.** −5, −2, 2, 5 **5.** −3, $-i$, $i$, 3 **6.** −4, −2, 2, 4 **7.** 25
**8.** 16 **9.** 1, 16 **10.** −2, $-2i\sqrt{2}$, $2i\sqrt{2}$, 2 **11.** −3, −2, 2, 3 **12.** $-\sqrt{3}$, $-2i$, $2i$, $\sqrt{3}$
**13.** −2, $-\sqrt{2}$, $\sqrt{2}$, 2 **14.** −1, −i, $i$, 1 **15.** −2, $-2i$, $2i$, 2 **16.** $-2\sqrt{2}$, $-2\sqrt{2}i$, $2\sqrt{2}i$, $2\sqrt{2}$
**17.** −3, $-\sqrt{7}$, $\sqrt{7}$, 3 **18.** 4 and 9 **19.** 4 and 49 **20.** 1 and 64 **21.** 1 and 64 **22.** −2, $-\frac{1}{2}$, $\frac{1}{2}$, 2
**23.** $-\frac{1}{2}$ and $\frac{1}{2}$ **24.** −1 and 1

## Objective 9.3.2

**1.** 4 **2.** 5 **3.** 5 **4.** 16 **5.** 1 and 4 **6.** 9 **7.** 3
**8.** 3 **9.** 4 **10.** 16 **11.** 3 and 4 **12.** 5 **13.** 8 **14.** 3
**15.** 7 **16.** −3 **17.** −5 and 3 **18.** 4 **19.** 9 **20.** 2 **21.** No solution
**22.** 2 and 6 **23.** −3 and 1

## Objective 9.3.3

**1.** −1 and 6 **2.** 3 and 4 **3.** −1 and 4 **4.** 1 and 6 **5.** −5 and 3 **6.** $-\frac{21}{4}$ and −1 **7.** $-\frac{1}{3}$
**8.** $-\frac{1}{5}$ and 5 **9.** 4 and 10 **10.** −1 and 7 **11.** −4 and $-\frac{9}{4}$ **12.** 0 and 4 **13.** −2 and $-\frac{2}{9}$ **14.** −4 and $\frac{3}{5}$
**15.** −3 and $-\frac{1}{2}$ **16.** −3 and 19 **17.** $-\frac{3}{5}$ **18.** −6 and 3 **19.** $-\frac{5}{2}$ and −1

## Objective 9.4.1

**1.** 10 ft and 4 ft **2.** 5 m, 8 m **3.** 5 and 6 **4.** −5 and −3 or 3 and 5 **5.** 2 **6.** −9 and −7 or 7 and 9 **7.** 9 h
**8.** smaller: 24 min; larger: 8 min **9.** older: 34 min; newer: 30 min **10.** 1st car: 48 mph; 2nd car: 60 mph

## Objective 9.5.1

**1.** 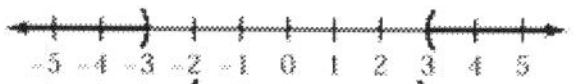


**2.** 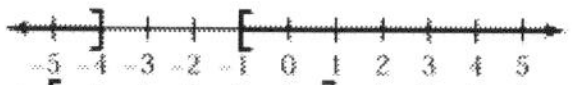


**3.** 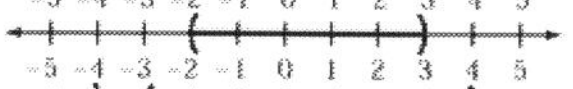


**4.** 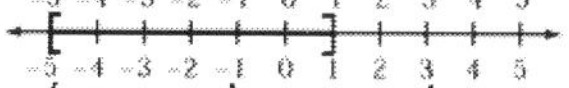


**5.** 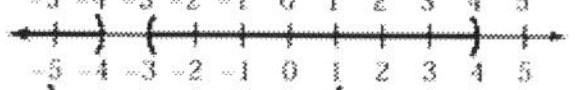


**6.** 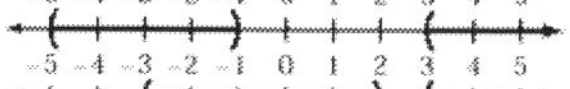


**7.** 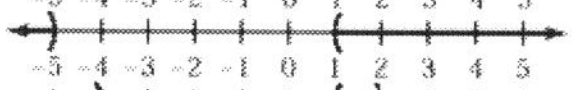


**8.** 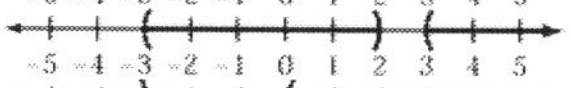


**9.** 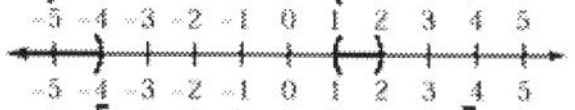


**10.** 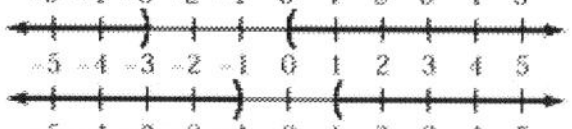


**11.** 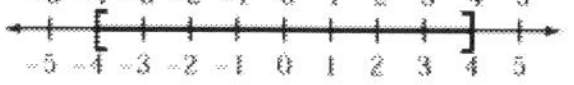


**12.** 

# Answers to AIM for Success Practice Sheets

## Objective 9.6.1

**1.** Vertex: $(0, -1)$
Axis of symmetry: $x = 0$

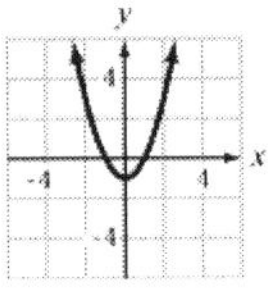

**2.** Vertex: $(0, 2)$
Axis of symmetry: $x = 0$

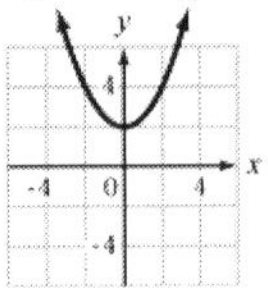

**3.** Vertex: $\left(-\frac{3}{2}, -\frac{9}{4}\right)$
Axis of symmetry: $x = -\frac{3}{2}$

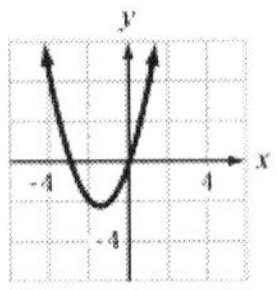

**4.** Vertex: $\left(\frac{3}{2}, \frac{1}{4}\right)$
Axis of symmetry: $x = \frac{3}{2}$

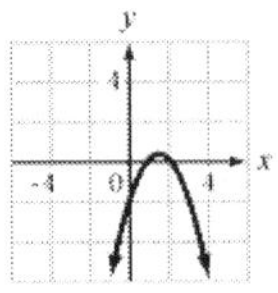

**5.** Domain: $\{x \mid x \in \text{real numbers}\}$
Range: $\left\{y \middle| y \geq -\frac{9}{4}\right\}$

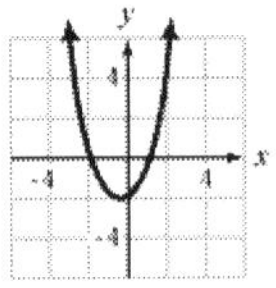

**6.** Domain: $\{x \mid x \in \text{real numbers}\}$
Range: $\{y \mid y \leq 1\}$

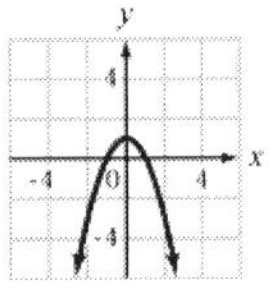

**7.** Domain: $\{x \mid x \in \text{real numbers}\}$
Range: $\left\{y \middle| y \geq -\frac{25}{8}\right\}$

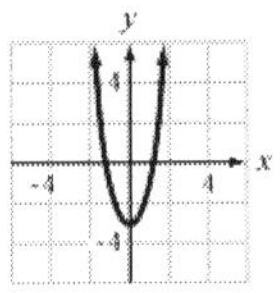

**8.** Domain: $\{x \mid x \in \text{real numbers}\}$
Range: $\left\{y \middle| y \geq -\frac{9}{4}\right\}$

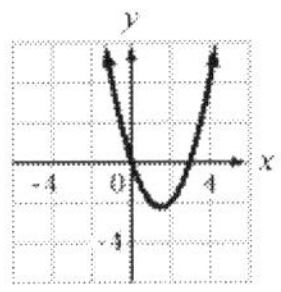

## Objective 9.6.2

**1.** $(-1, 0), (1, 0)$ **2.** $(-4, 0), (4, 0)$ **3.** $(0, 0), (1, 0)$ **4.** $(0, 0), (2, 0)$ **5.** $\left(-\frac{1}{2}, 0\right), (2, 0)$ **6.** $(-6, 0), (1, 0)$

**7.** $\frac{1}{3}, 1$ **8.** $-3, 4$ **9.** No zeros **10.** No zeros **11.** $-1, 4$ **12.** $-3, 5$ **13.** two

**14.** No $x$-intercepts **15.** two **16.** No $x$-intercepts **17.** two **18.** two

**19.** No $x$-intercepts **20.** one **21.** one **22.** two **23.** two **24.** No $x$-intercepts

## Objective 9.7.1

**1.** minimum: $-1$ **2.** minimum: $-4$ **3.** maximum: 13 **4.** maximum: 2 **5.** minimum: $-11$

**6.** minimum: $-\frac{1}{4}$ **7.** maximum: 11 **8.** minimum: $-7$ **9.** maximum: $\frac{25}{4}$ **10.** minimum: $-25$

**11.** minimum: $-\frac{11}{2}$ **12.** maximum: 1 **13.** minimum: $-3$ **14.** minimum: $\frac{15}{4}$

**15.** maximum: $\frac{22}{3}$ **16.** maximum: $\frac{17}{8}$

## Objective 9.7.2

**1.** 144 ft **2.** 131 ft **3.** 5 days **4.** \$187.50 **5.** 7 s **6.** 25 and 25 **7.** 15 and 15

**8.** $-10$ and 10 **9.** 14 ft by 14 ft; 196 $ft^2$ **10.** 8 ft by 8 ft; 64 $ft^2$

# CHAPTER 10

## Objective 10.1.1

1. 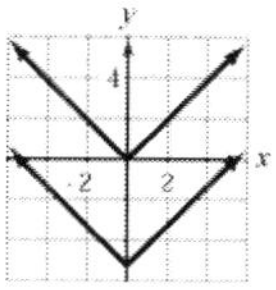

2. 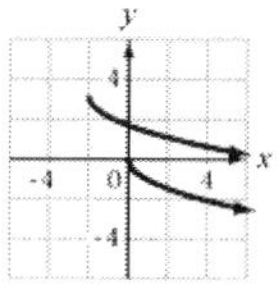

3. 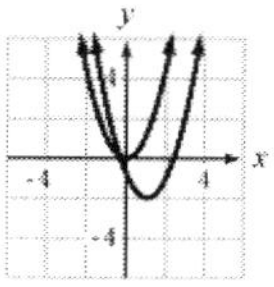

4. 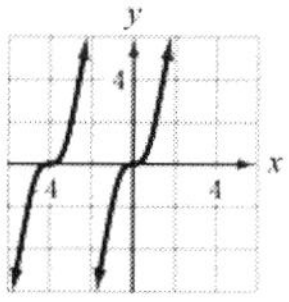

5. 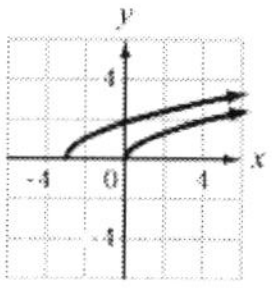

6. 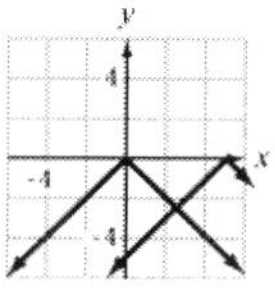

7. 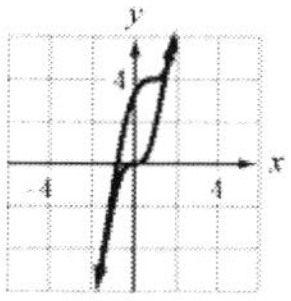

8. 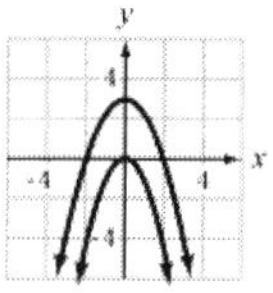

## Objective 10.2.1

**1.** –8 **2.** 10 **3.** –11 **4.** 11 **5.** $\frac{1}{2}$ **6.** –90 **7.** 89

**8.** 3 **9.** 7 **10.** –242 **11.** undefined **12.** $\frac{5}{6}$ **13.** –1 **14.** 36

**15.** 12 **16.** $-\frac{5}{8}$

## Objective 10.2.2

**1.** 41 **2.** 9 **3.** 51 **4.** –21 **5.** $10x - 1$ **6.** $10x + 31$ **7.** 11

**8.** $x^2 - 6x + 14$ **9.** $x^2 + 2$ **10.** 14 **11.** –25 **12.** –1461 **13.** 47 **14.** $8x^3 - 17$

**15.** $128x^3 - 480x^2 + 600x - 253$ **16.** –17

## Objective 10.3.1

**1.** No **2.** Yes **3.** Yes **4.** No **5.** No **6.** No **7.** No

**8.** No

## Objective 10.3.2

**1.** $f^{-1}(x) = \frac{1}{4}x - 2$ **2.** $f^{-1}(x) = x - 5$ **3.** $f^{-1}(x) = \frac{1}{3}x + 2$ **4.** $f^{-1}(x) = -x + 4$

**5.** $f^{-1}(x) = -4x + 8$ **6.** $f^{-1}(x) = -\frac{1}{2}x + 3$ **7.** $f^{-1}(x) = \frac{3}{2}x + 3$ **8.** $f^{-1}(x) = 4x + 24$

**9.** $f^{-1}(x) = \frac{1}{3}x - \frac{2}{3}$ **10.** $f^{-1}(x) = \frac{1}{5}x - \frac{1}{5}$ **11.** $f^{-1}(x) = -\frac{1}{4}x + \frac{1}{2}$ **12.** $f^{-1}(x) = \frac{1}{4}x - \frac{3}{4}$

**13.** Yes **14.** Yes **15.** No **16.** No

# CHAPTER 11

## Objective 11.1.1

**1.** 4 **2.** $\frac{1}{4}$ **3.** 1 **4.** 16 **5.** $\frac{1}{16}$ **6.** 64 **7.** 3

**8.** 9 **9.** $\frac{1}{3}$ **10.** 1 **11.** $\frac{1}{9}$ **12.** 9 **13.** $\frac{1}{3}$ **14.** 3

**15.** $\frac{1}{81}$ **16.** 1 **17.** 3 **18.** 81 **19.** 3 **20.** 81 **21.** 19.683

## Objective 11.1.2

**1.** 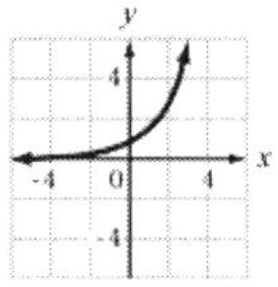 **2.** 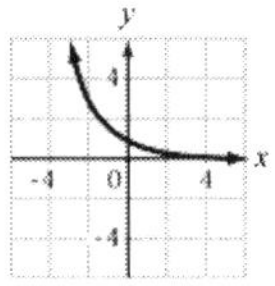 **3.** 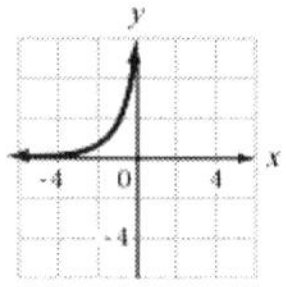 **4.** 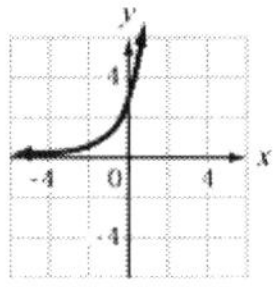

**5.** 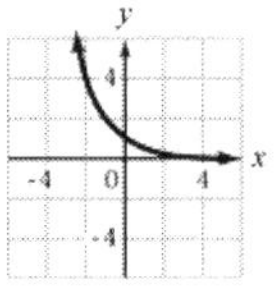 **6.** 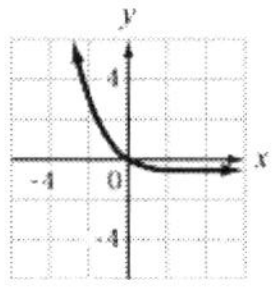 **7.** 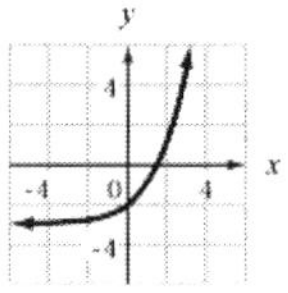 **8.** 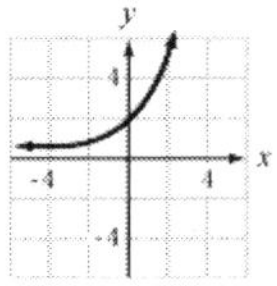

## Objective 11.2.1

**1.** $\log_4 64 = 3$ **2.** $\log_7 49 = 2$ **3.** $\log_{64} \frac{1}{8} = -2$ **4.** $\log_{10} 10{,}000 = 4$ **5.** $\log_4 1 = 0$ **6.** $\log_p q = x$

**7.** $2^0 = 1$ **8.** $2^{-2} = \frac{1}{4}$ **9.** $e^c = x$ **10.** 2 **11.** –4 **12.** 2 **13.** 4

**14.** 4 **15.** 0 **16.** 64 **17.** 16 **18.** 216 **19.** 1 **20.** 10,000
**21.** 125

## Objective 11.2.2

**1.** $\log_2 x^2 y$ **2.** $\log_5 \frac{x^3}{y^4}$ **3.** $\log_5 x^4$ **4.** $\log_6 \frac{1}{x^3}$ **5.** $\log_6\left(\frac{x^3}{y^4}\right)$ **6.** $\log_3 \frac{x^3 z^3}{y^2}$ **7.** $\log_6 \sqrt[3]{\frac{x}{z}}$

**8.** $\log_3 \frac{x^5 z^3}{y^3}$ **9.** $\log_5 \sqrt{\frac{x^5 z}{y^3}}$ **10.** $\log_3 2 + \log_3 x$ **11.** $3\log_2 x$ **12.** $3\log_5 v - 7\log_5 x$

**13.** $\ln x + 3\ln y + 2\ln z$ **14.** $\frac{2}{3}\log_6 r + \frac{1}{3}\log_6 w$ **15.** $\frac{1}{2}(2\log_5 x - \log_5 y)$ **16.** 1.2920 **17.** 1.7924

**18.** 1.8981 **19.** 0.7326 **20.** 0.9692 **21.** 0.0937

## Objective 11.3.1

**1.** 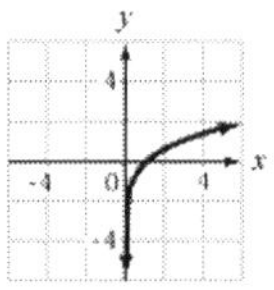 **2.** 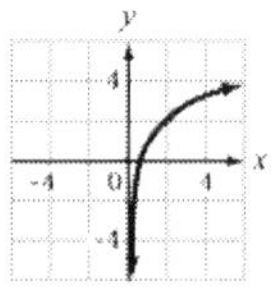 **3.** 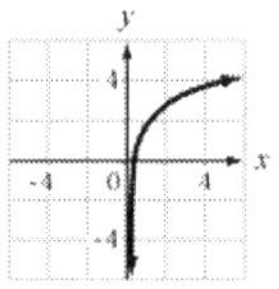 **4.** 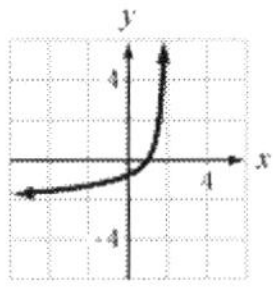

**5.** 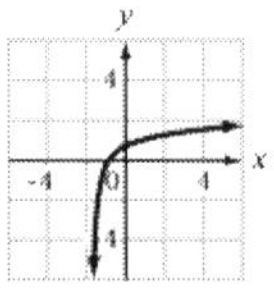 **6.** 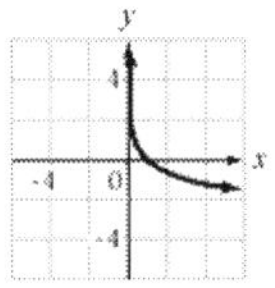 **7.** 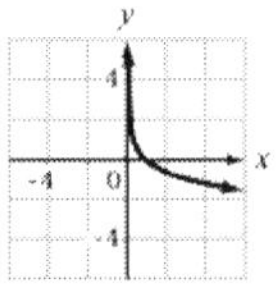 **8.** 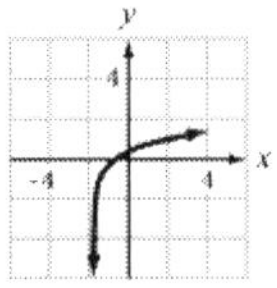

## Objective 11.4.1

**1.** –2 **2.** 1 **3.** 5 **4.** –4 **5.** 0.7782 **6.** 0.8842 **7.** –2
**8.** –4 **9.** 3.2656 **10.** 1.7567 **11.** 1.3802 **12.** 1.5798 **13.** –3 **14.** –2.4651

**15.** 4.1701 **16.** 0.7713 **17.** 2 **18.** –0.4651 **19.** $\frac{2}{3}$ **20.** 1 **21.** 3

**22.** –2 **23.** 3 **24.** –4

## Objective 11.4.2

**1.** 24 **2.** 5 **3.** 14 **4.** 26 **5.** 5 **6.** 42 **7.** 5

**8.** –8 and 2 **9.** –4 and 9 **10.** 1 **11.** $\frac{5}{2}$ **12.** 3 **13.** No solution **14.** $\frac{3+\sqrt{41}}{4}$

**15.** 2 **16.** $-\frac{2}{3}$ and 3

## Objective 11.5.1

**1.** \$12,751 **2.** 6 years **3.** 68% **4.** 10th week **5.** 0.4 cm **6.** 75% **7.** 398 lines
**8.** 63,100 times

# CHAPTER 12

## Objective 12.1.1

**1.** Vertex: $\left(\frac{1}{2}, -\frac{9}{4}\right)$; Axis of symmetry: $x = \frac{1}{2}$

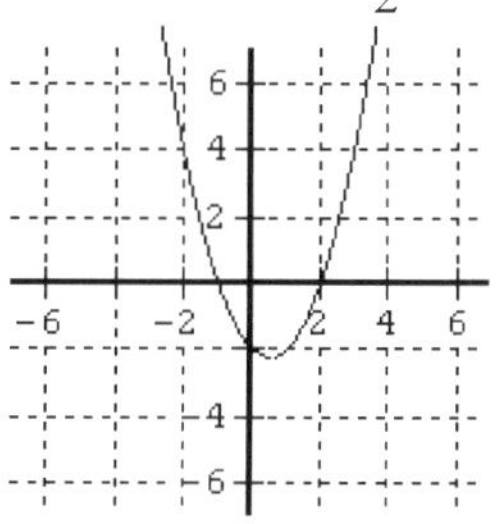

**2.** Vertex: (–1, –4); Axis of symmetry: $x = -1$

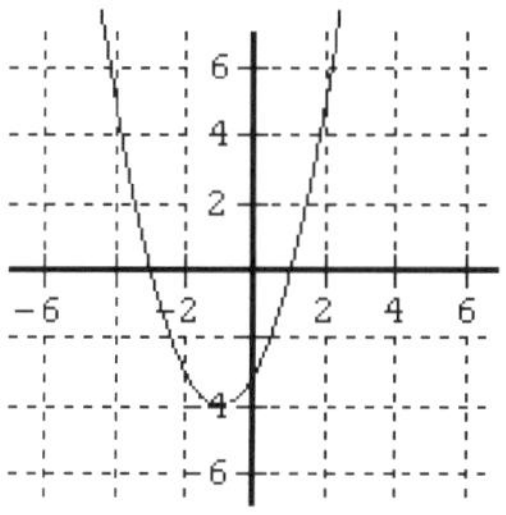

**3.** Vertex: (1, 5); Axis of symmetry: $x = 1$

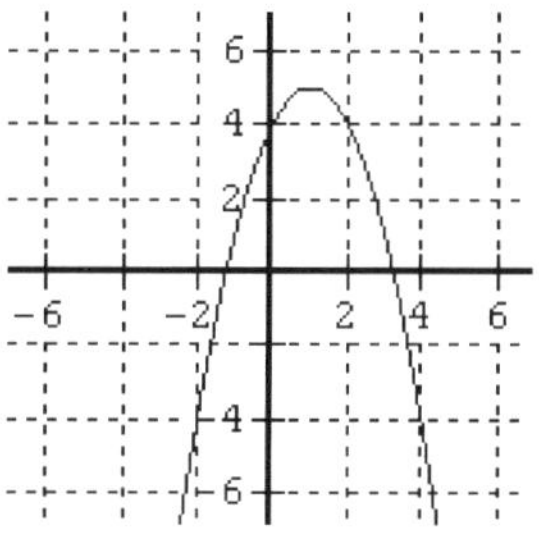

**4.** Vertex: (3, 4); Axis of symmetry: $x = 3$

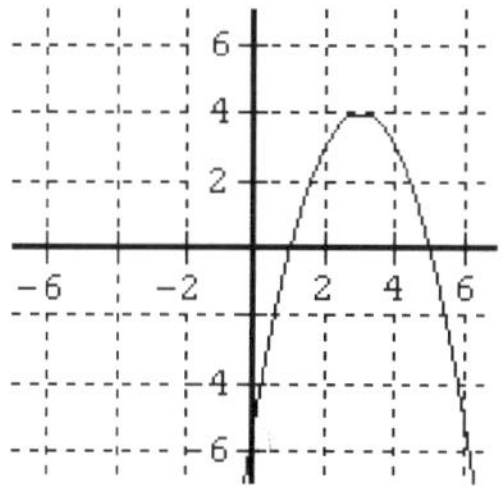

**5.** Vertex: (2, –1); Axis of symmetry: $y = -1$

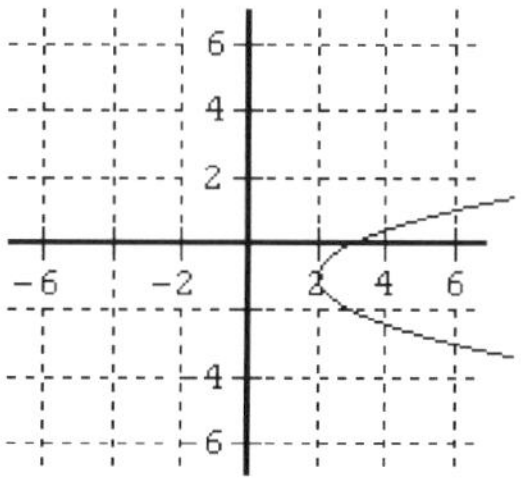

**6.** Vertex: (–3, 0); Axis of symmetry $y = 0$

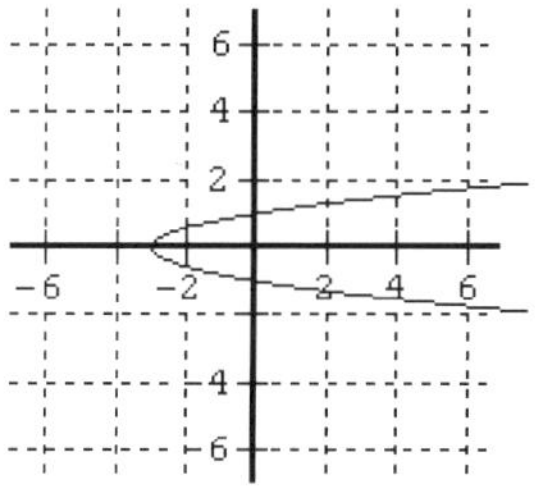

**7.** Vertex: (4, 0); Axis of symmetry: $y = 0$

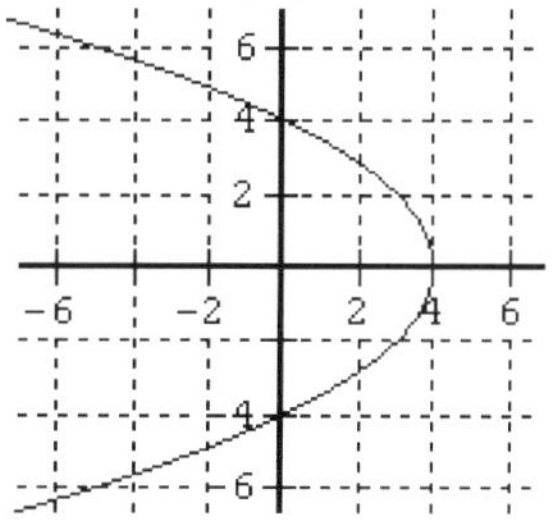

**8.** Vertex: (2, 0); Axis of symmetry: $y = 0$

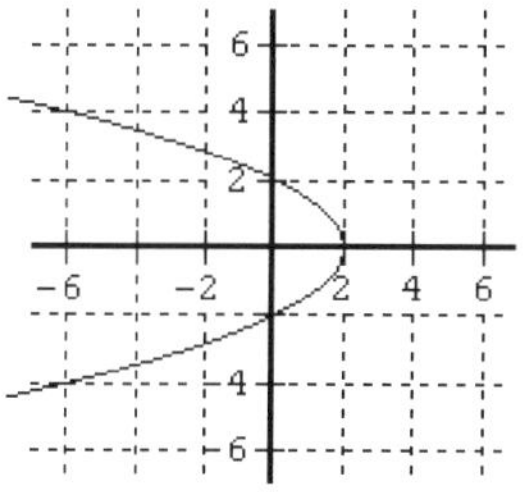

## Objective 12.2.1

**1.**

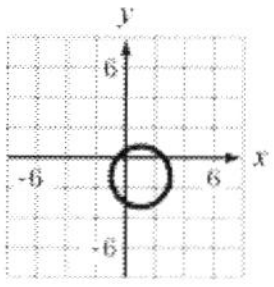

**2.**

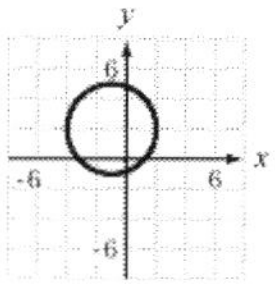

**3.**

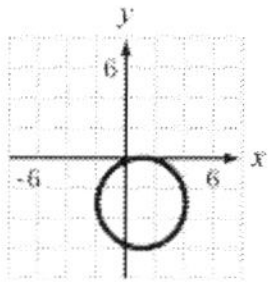

**4.**

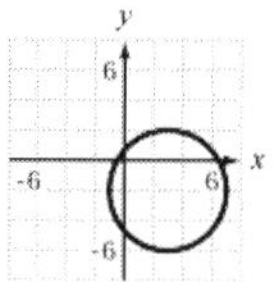

**5.** $x^2+(y+4)^2=16$

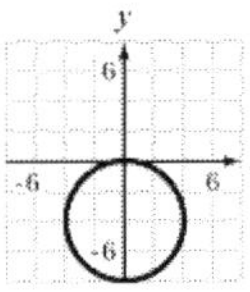

**6.** $(x+2)^2+(y+1)^2=4$

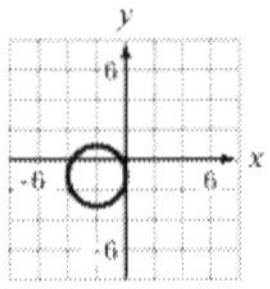

**7.** $(x+1)^2+(y-3)^2=25$

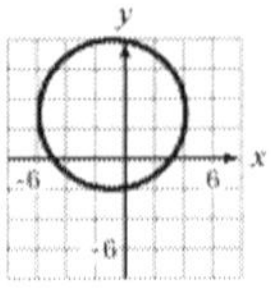

**8.** $(x-2)^2+(y+3)^2=16$

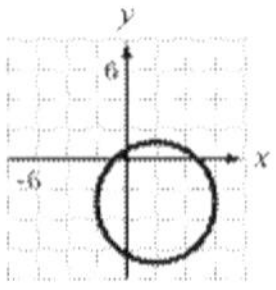

## Objective 12.2.2

**1.** $(x-5)^2+(y+1)^2=4$

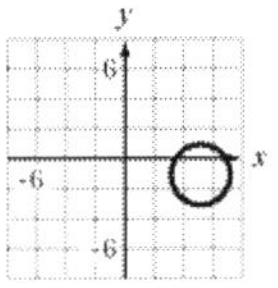

**2.** $(x+4)^2+(y-2)^2=16$

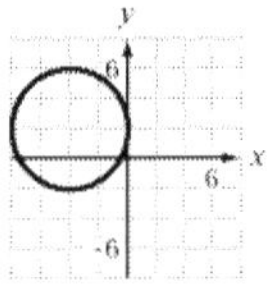

**3.** $(x-5)^2+(y+2)^2=\frac{25}{4}$

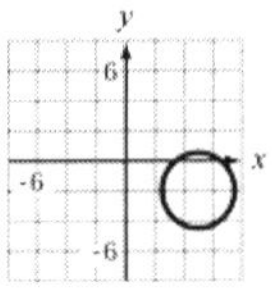

**4.** $(x+3)^2+(y-5)^2=25$

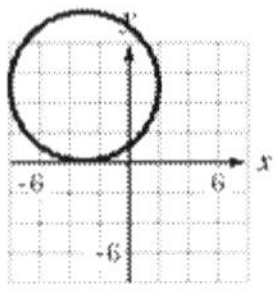

**5.** $(x-3)^2+(y+4)^2=9$

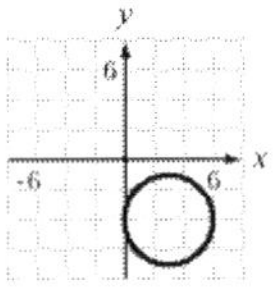

**6.** $(x-3)^2+(y-2)^2=4$

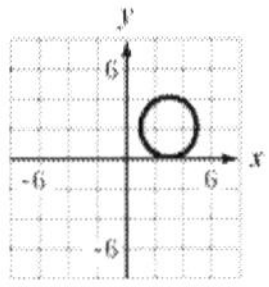

**7.** $(x+3)^2+(y-2)^2=16$

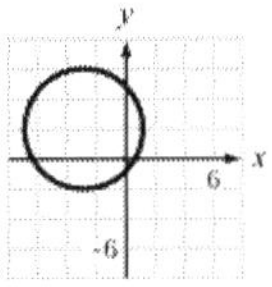

**8.** $(x+3)^2+(y-1)^2=4$

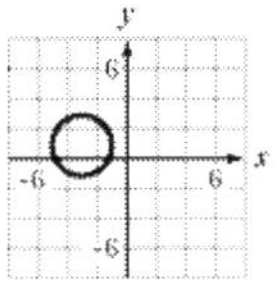

## Objective 12.3.1

**1.**

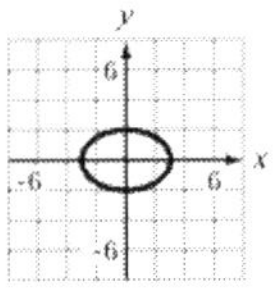

**2.**

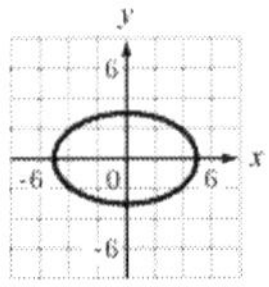

**3.**

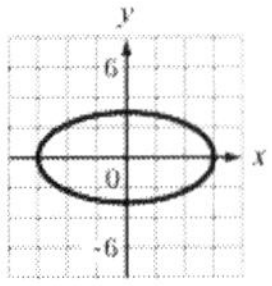

**4.**

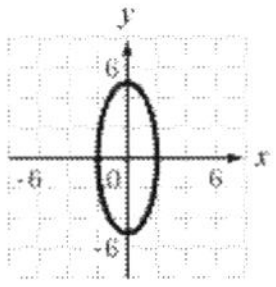

**5.**

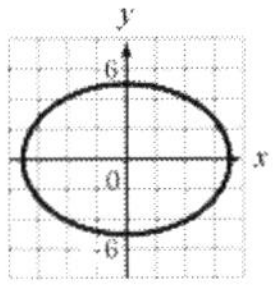

**6.**

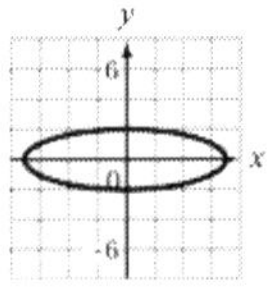

**7.**

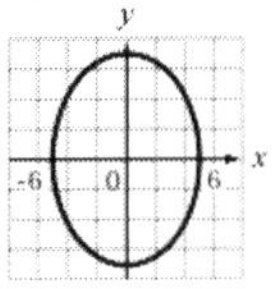

**8.**

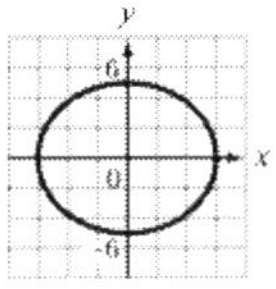

## Objective 12.3.2

**1.**

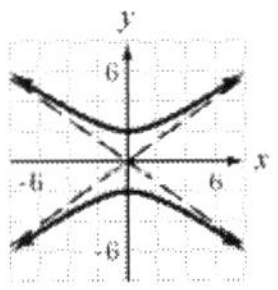

**2.**

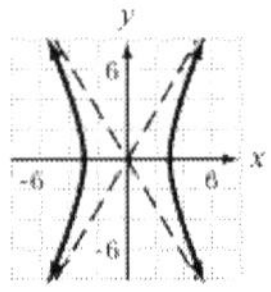

**3.**

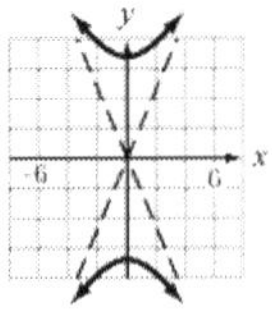

**4.**

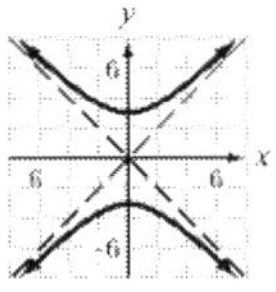

**5.**

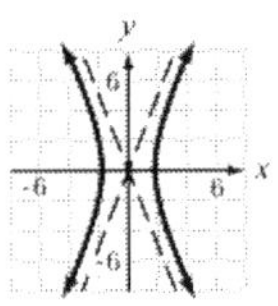

**6.**

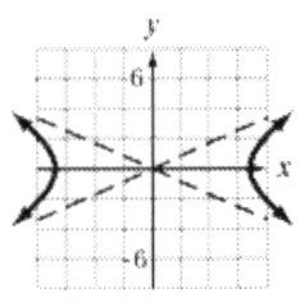

**7.**

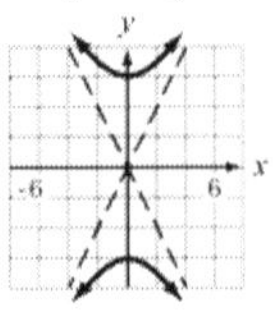

**8.**

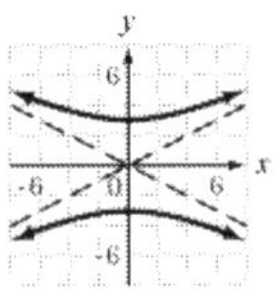

## Objective 12.4.1

**1.** (6, 14) and (–1, 0) **2.** (–2, 1) and (7, 19) **3.** (–5, –13) and (3, 3) **4.** (0, 0) and (7, 14)

**5.** (–1, –1) and (2, 2) **6.** (–3, –1) and $\left(-2,-\frac{3}{2}\right)$ **7.** (1, 3) and (3, 1) **8.** $\left(-\frac{7}{5},-\frac{24}{5}\right)$ and (3, 4)

**9.** (–2, –3) and (3, 2) **10.** $\left(-\frac{8}{3},\frac{1}{3}\right)$ and (4, 3) **11.** (–2, 3) **12.** (–6, –3) and (–6, 3), (6, –3) and (6, 3)

**13.** (–2, 0) and (2, 0) **14.** (–3, –2) and (–3, 2), (3, –2) and (3, 2) **15.** (–1, 2) and $\left(-\frac{1}{9}, \frac{22}{9}\right)$
**16.** (–1, –2) and (–1, 2), (1, –2) and (1, 2)

Made in United States
Orlando, FL
04 November 2026

90200934R00109